S0-BSQ-655

Nat hist
20-
17476

Farm

Farm

Grant Heilman

Abbeville Press · Publishers · New York

To Barbara

A Roundtable Press Book
Directors: Susan E. Meyer, Marsha Melnick
Editor: Virginia Croft
Assistant Editor: Marguerite Ross

Published by Abbeville Press
Project Editor: Walton Rawls
Art Direction: Renée Khatami
Designer: Otto H. Barz
Copy Chief: Robin James
Production Manager: Dana Cole

Library of Congress Cataloging-in-Publication Data

Heilman, Grant, 1919–
Farm.
Bibliography: p.
Includes index.
1. Agriculture—United States. 2. Farms—
United States. 3. Agriculture—United States—
Pictorial works. 4. Farms—United States—Pictorial
works. 5. United States—Description and travel—
1981– —Views. I. Title.
S441.H45   1988   630'.973   88-14657
ISBN 0-89659-889-6

Copyright © 1988 by Cross River Press, Ltd. All rights reserved under International and Pan-American Copyright Conventions. No part of this book may be reproduced or utilized in any form or by any means, electronic or mechanical, including photocopying, recording, or by any information storage and retrieval system, without permission in writing from the publisher. Inquiries should be addressed to Abbeville Press, Inc., 488 Madison Avenue, New York, N.Y. 10022. Printed and bound in Japan. First edition.

Contents

Animal Agriculture   145

Vegetables, Fruits, and Nuts   199

The Changing Scene   255

Preface

In 1900 one farm worker supported himself and one city worker. Today one farm worker supports more than eighty city dwellers. As a result, most of us know more about the African veldt and its inhabitants than we know about the American farm and the American farmer.

This book is an attempt to remedy that. While it is intended primarily for the non-farmer, I'm hopeful that farmers can enjoy it also, for the truth is that farmers know little about one another. The difference between a dairy farmer in Vermont's Connecticut Valley and a carrot grower in California's Imperial Valley is as great as the difference between a resident of Manhattan and a small town citizen in Iowa.

In the 1920s and 1930s American politics was faced with what passed as the "Farm Bloc." It was assumed that all farmers had similar interests, unlike the urban nonfarmers, and therefore would routinely vote alike, creating this so-called bloc.

Today that's seldom the case. As the farmer has become more specialized, the Farm Bloc has disintegrated. Corn farmers want high corn prices; beef cattle feeders want low corn prices. California citrus growers haven't any reason to back high—or low—dairy prices, but since citrus growers are also consumers, they naturally favor cheap milk—but not cheap oranges. Dairy farmers favor cheap oranges, of course.

As a percentage of the total population, the farmer has almost disappeared. Although this depends a little on the definition of what a farmer is (and there are plenty of varied definitions), nevertheless the farmer is no longer a big factor in our total population. The agricultural census counted nearly 2.4

million farms in 1985, but its total was based on a very broad definition: "anyone who sells or could sell more than $1,000 of farm products during the year." A more realistic estimate is that there are fewer than 1 million farmers who make their full living from farming.

This is not to say that because there are fewer farmers they are any less important to us. The farmer may have less political clout than in the past, but bear in mind a bumper sticker making the rounds in the Midwest: "Don't criticize a farmer with your mouth full!" Farmers keep our mouths full and our tables overflowing with reasonably priced products and are responsible for the employment of many Americans who are not farmers themselves. This creates clout.

Farming claims to be America's largest industry and the country's largest employer. There may be fewer than 1 million actual farms, but the farming industry employs perhaps 21 million people, including those who manufacture farm machinery, farm chemicals, and other farm inputs and those who process farm products into finished products for the grocer's shelves. Looked at that way, it's by far our largest industry.

I've included here as little as necessary about the politics of agriculture. In a chapter at the end of the book, I've attempted to summarize the political issues, and wherever I felt it was necessary, I've discussed specific governmental programs in connection with specific crops. Farming is our largest industry, full of problems that have been overcome and a lot that haven't yet been solved. It's an industry and at the same time a way of life. My hope is that this book will give the reader some insight into that way of life.

Grant Heilman
November 1987

Necessary Facts

Some plain facts and figures are reviewed here to further your understanding of agricultural America. You may want to keep your thumb in this section so you can refer back to it.

In the preface I mentioned there are several definitions for the word *farm*. According to the current agricultural census, a place that sells a minimum of $1,000 of agricultural products in a year qualifies as a farm. (This incredibly low figure has likely caused more confusion about farm economics than any other single fact!) When we say we had 2,214,000 farms in 1986, remember that more than 60 percent of these had annual gross sales of less than $20,000, and it's pretty hard to make a living from a farm with sales that low. As you can tell, the census figures confuse the farming issue if they are not adequately explained.

A couple of semantics problems are also confusing. The first one is international. In the United States and Canada, what we call wheat is a specific series of plants of the genus *Triticum*. Corn is a totally different plant, of the genus *Zea*. In other English-speaking countries, *corn* can mean simply the seed of whatever cereal crop is important to a particular region. The word is used to refer to small grains, including wheat but also including what we call barley and what we call oats. What we in the United States call corn is known elsewhere as maize. Some versions of the Bible refer to corn but actually mean small grains, usually but not always wheat. The King James version of the Bible uses both the words *wheat* and *corn* as translations for the same Hebrew word. The plant we call corn was unknown in biblical times—at least in biblical lands.

Not to be confusing, but in the United States and Canada, sorghum—a crop slightly similar to corn—is known in some places as milo, in others as sorghum, and in still a few others as maize. The general tendency is to settle on *sorghum*, which is the term I've used, although *milo* is still in very common usage.

The measurements used in farm talk often don't mean much to city dwellers but are vital to farmers. An acre is 43,560 square feet; that's a little more than 200 by 200 feet. A city lot, 100 by 200 feet, is slightly less than half an acre. A football field without end zones is 48,000 square feet, about 10 percent larger than an acre.

A section is 640 acres, which is also a square mile. Much of the country was originally parceled out to buyers in sections— square miles—or in quarter sections, 160 acres. Measurement still works that way. "I farm 10 sections of wheat up in eastern Wyoming" means 10 square miles.

Crop production is usually measured in bushels per acre, as in "He got 150 bushels to the acre, a really good corn crop." A bushel is 2,219 cubic inches—actually a cubic measurement. Practically speaking, however, bushels are bags filled more or less to the same level near the top. A bushel usually weighs 60 pounds, although shelled corn is 56 pounds, wheat is recorded as 60 pounds, and rice is 45 pounds. But pounds are seldom used; bushels are.

Water used for irrigation is "put on" in numbers of inches. "We put on 6 inches of water" is like "We had 6 inches of rain." In cubic measure, water is talked of in acre feet. An acre foot is water a foot deep covering an acre of ground. The measurement of reservoirs is usually in acre feet. The flow of rivers is usually in cubic feet per second; the Mississippi averages 640,000 cubic feet per second (cfs) at the mouth. When I refer to precipitation, incidentally, I include snowfall as well as rainfall. Precipitation is usually figured at 10 inches of snow to 1 inch of water. Snow cover is necessary not only to supply moisture but also to prevent wind damage to winter wheat.

The American Farm

It seems to be widely believed that the American family farm is something that disappeared at some distant time in the past, along with Model T Fords and suspenders. Untrue. Along with this belief usually goes the idea that giant corporations have taken over all farming and that much of this country's farmland has been gobbled up by foreigners. Also untrue.

What is really happening to the family farm? The number of family-owned farms certainly has decreased drastically. First, remember the definition of a farm, according to the agricultural census: a place that sells a minimum of $1,000 in agricultural products annually, and all figures are skewed by that. If you've got a horse in the backyard and sell her for $1,100 and it's an agricultural census year, you're a farmer and are counted among the 2.21 million farms.

In 1986, of the total 2.21 million farms, half grossed less than $10,000 in sales, but these small farms occupied only 11 percent of the billion acres classified as farmland. About 14 percent of the total farms had sales of more than $100,000, but these bigger farms owned half of the land.

So, what about that endangered species, the family farm? We keep hearing about its demise on television and reading about it in magazine articles. Certainly the number of farms has dropped drastically. The year we had the most farms was 1935, when there were 6.8 million. By 1986 we were down to 2.21 million—a drop of two-thirds—and the trend is likely to continue. But look at the number of acres farmed. We farmed the highest number of acres in 1954, 1.2 billion, but by 1986 the acreage farmed had dropped only 16 percent. Obviously, each farm had become a lot larger; the average farm size went from 252 acres in 1954 to 455 acres in 1986. Did huge corporations buy up all the small family farms?

Less than one-half of 1 percent of the farms are owned by nonfamily corporations, but they control a much higher percentage of the land. Is the family farm alive and healthy? Has the media been bamboozling the public? Not exactly. The big corporation intruding into the farmland has become the whipping boy for the problems of the family farm, and the family farm has decreased greatly in numbers, but there is not necessarily a cause-and-effect relationship.

Mostly small farms became larger by buying up neighboring farms, which simply disappeared as entities. In some cases, of course, big corporations did buy up existing small

In fifty years the number of individual farms in the United States has dropped by two-thirds, and the decrease in farm numbers is likely to continue as the number of acres per farm continues to increase. Only about 14 percent of our farms had sales greater than $100,000, but this 14 percent owned half of the total farmland.

farms, but corporate ownership has not greatly expanded in terms of the number of acres owned. A farm ownership survey done in 1946 shows about the same concentration as there is now. The largest 1 percent of owners holds almost 30 percent of the acreage—and has for a long time.

In terms of dollars in sales, the biggest farms have expanded rapidly. This is partly a question of semantics, for we quickly get into what is farming and what is agribusiness. For example, it is thought that of the ten major farm owners, seven are in poultry. Whether it's fair to consider the total sales figures of poultry producers as agricultural, which is often done, is questionable, since these sales figures include what is traditionally nonfarm income.

Vertical integration is the reason for the confusion. It refers to ownership of more than one stage of farming and marketing: production, processing, packaging, wholesaling, and sometimes retailing. So far vertical integration has been limited to a relatively small number of farm products. It hasn't yet much reached the blockbuster crops. It's come rapidly and completely into poultry, partially into beef, is trying to get going in hogs, and is active in a few specialty crops. But the big grain companies (many of which are privately owned, and a number of which are foreign owned) that control the buying and selling of grain haven't yet become a big factor in actually growing the grain. They figure it's cheaper to buy it and trade it than to grow it, and they are probably right.

If farmland prices continue to plummet—during the 1980s they've dropped by as much as 30 to 50 percent—more big corporations

Abandoned farmhouses are common, particularly through the Midwest. Abandoned farmland, however, is almost nonexistent. While the number of farms has plummeted, the acreage farmed has dropped only 16 percent in the past thirty years. Much of the loss of farmland has been caused by urban growth absorbing farmland near cities.

could conceivably become interested in putting their capital into farming operations. The low profit margin from growing farm products, however, particularly the blockbuster crops, will probably dissuade corporations from becoming farmers. Real estate companies may be tempted, and some are already in the farming business, but their interests will continue to be primarily in development, only secondarily in agriculture.

Guessing at the future, I would not look for total corporate ownership or total demise of the family farm. I would look for some increase in vertical integration, particularly in animal production, but also the continuance of family farms, smaller in number but larger in size.

It's remarkable how frequently the following conversation takes place when I'm visiting with farmers: "Things are terrible—I lose money every year. There's got to be some better way of making a living. We're being badly treated by the government programs, and I'll be lucky to be able to hold out for another year."

I shake my head in sympathy during this tirade and know what's going to come later in the conversation: "Charlie, he owns the farm next to mine over in the northwest corner there. He's really in trouble, and when it gets a little worse, well, he'd like to retire and get off the farm, and if he'd just drop his asking price I could arrange the financing to buy his place. It'd fit into my operation slick as can be, not take any more equipment, and I could manage the labor. He's gettin' his price closer to being able to cash flow it. I'll just have to wait him out."

I believe that many of these dreams of expansion will take place. The family farm is going to continue: fewer in number, bigger in acreage. It's going to be hemmed in by increasing government restrictions, by estate problems, by financial squeezes, by the contractual pressure of increased integration in the food industry. However, I doubt we will ever see the day of total corporate or governmental farming. Farming has evolved to this point and should be able to avoid revolution and continue its evolution.

A number of states, alarmed at the possible intrusion of corporate farming, have passed laws to restrict farm ownership to "family farming." The efficacy of these laws remains to be seen, as well as their legality. Effectively defining *family farming* and *corporate farming* when the two are approaching each other is difficult, particularly in a politically charged atmosphere.

If private ownership is continuing, then what is owned, what is rented, and who owns and rents? Of the total farmland, 60 percent is owned by the persons farming it, 40 percent is rental ground. About 34 percent of the total land in farms is owned by nonfarmers, but most of these owners live nearby. Only 6 percent of the farmland is owned by persons who do not live in the same state as the farm they own.

The majority of farm owners are middle-aged or older: about 30 percent are over sixty-five, while only 6 percent are younger than thirty-five. Minorities own very little farmland. Blacks and Hispanics each own less than 1 percent of the totals. The position of women in ownership is not very clear, because husband-wife holdings and family partnerships are not subject to easy quantifying. Although tax law changes are tending to benefit transferring farm ownership from

The greatest decrease in farm numbers is due to the increase in farm size. This has been accomplished mostly by increased mechanization of crop handling. The effects are most noticeable in the grain belt of the Midwest, where large fields and flat terrain lend themselves to mechanized farming.

husband to wife, it is usually thought that about 85 percent of the noncorporate farmland is still owned by males.

Because of the peculiar definition of a farm used by the Census of Agriculture, 63 percent of our farmers sell less than $20,000 worth of farm products each year but account for less than 10 percent of the agricultural production. These farmers, and there are almost 1.4 million of them, have on the average a negative farm income—they lose money from their farming. They are thus dependent on nonfarm employment for their living.

There are no statistics that tell how many of these part-timers are trying to make money from farming and how many are hobby farmers, simply having an enjoyable time. For many years I was happy to be classified as a farmer because I grew and sold Christmas trees beyond the $250 annual sales then required to qualify as a farmer. I was a happy hobbyist.

Including the large group that doesn't bother to sell any products at all, as well as the group that sells but loses money, there are millions of us having fun growing things. An entire network of services has sprung up to help us. Special publications supply us with information, and their luscious advertising tempts us with products specially designed for the small operator. Local colleges offer us courses, seed companies pander to us, 4-H programs engulf our kids, Extension Service personnel are delegated to answer our questions. Nevertheless, when the figures are in, we've still contributed less than 10 percent of the agricultural production. But we have had fun.

There are other part-timers, however, who make a more major contribution to agricul-

The increase in mechanization and in the size of equipment means the farmer can handle more land, but it also means the farmer needs a much higher capital investment. Tractors that cost $50,000 to $100,000 are not uncommon.

tural output and who usually show a profit at the end of the year. Much chicken raising, for example, has been taken over by part-timers, predominantly financed by the large integrated producers or by feed companies eager to increase sales. Automatic equipment in a broiler house means high production with little labor. A part-timer can easily raise more than a hundred thousand chickens a year. Other animals and some specialty crops also provide part-time opportunities that can be profitable, but it is more difficult to show a profit on part-time growing of major crops.

As work hours in industry continue to shorten and farming continues to mechanize, the number of part-time farmers is likely to increase. They will be able to replace the smallest full-time farmers if the full-timers do not expand their acreage to make more valuable use of their time.

Although part-timers are able to work in agriculture partly because shorter working hours in nonfarm jobs leave them more time for working on the farm, I believe there is another factor that is less measurable. Farming may be hard work but it is generally enjoyable, and many people recognize the joys of farming even if they find little profit. Not only part-timers and hobby farmers but many full-time farmers feel this way. In fact,

Corporate ownership of farms is very limited—less than one half of 1 percent of the farms are owned by non-family corporations. They do, however, control much more land than their numbers indicate, and this is likely to increase. At the other end of the scale are part-time farmers, who raise little of the major crops such as wheat but are active in specialty crops and animal agriculture. There are likely to be more part-time farmers. The small to middle-sized full-time farmers are the ones who will be increasingly squeezed.

I believe it is one of the major factors in the entire farm problem. Independent farming is pleasurable, and farmers are willing—down to a certain level—to accept less for their services from farming than if they were in some other activity.

One family who typifies this feeling farms north of Scottsbluff, Nebraska. They are part-time wheat growers, probably the most unusual part-timers I've encountered, and I'm not sure I've ever met any others. I stopped when I saw a combine working its way through a rolling Nebraska wheat field and got into a conversation with the combine driver when he came over to unload.

"We are not exactly your typical farmers," he admitted. "My brother and I inherited this land, and we have always hated the thought of selling it. There's not enough acreage to make a living, so we can't afford to live here, but we farm it. The old homestead used to be up there by that tree, but there are no houses here now." There was only a single tree visible

on the horizon, along with a steel machinery shed across the road. I asked where they lived, since there was no house on the place.

"Denver," he said. Denver, I realized, was something more than 200 miles away. "I'm an airline pilot, and my brother runs a limousine rental service. We commute up here weekends with our kids to raise the wheat."

Intrigued, I asked how they commuted. He looked at me as if I had never been exposed to the everyday world. "Well, naturally," he said, "we fly. The plane's over there in the steel shed the equipment is housed in. We bulldozed a dirt runway out on the edge of a wheat field." Covered with dust and sweat, he climbed back onto the combine.

"Is it fun?" I asked.

"You'd better believe it," he replied.

"Do you make money?" I shouted.

He slammed the door, and I could see him laughing as he gunned the engine and rumbled back into the field.

Part-timing is hard work—demanding,

time-consuming, frustrating, but still fun. Here's to it! There's something pretty important in it for many of us—not all of us, just some of us—something that gives us pleasure from dirt on our hands, from being responsible for something that grows. If we are to understand farm numbers, we should understand that 63 percent of what the census calls farmers are really not in it for the profit but for the fun.

We haven't yet dealt with those scare headlines about foreigners who are buying up our land. Foreign ownership actually comes to a little less than 1 percent of the total U.S. ag-ricultural land, and the foreign-owned land is scattered. Maine has the largest number of acres owned by foreign persons—10 percent of the state's agricultural land—but 91 percent of that is held by three companies, two of which are Canadian and one of which is a United States company that is partly foreign owned. These companies are really in the forestry business, not in crop farming.

The best way to study farm ownership and farm operations is to take a slow trip across the United States, looking at farming and talking with farmers. It's an incredibly rich country, diverse in its people and its farms.

About 60 percent of the total farmland is owned by the people who farm it, and 40 percent is rented. Farm owners tend to be middle-aged or older—almost a third of the farmers are over the age of sixty-five. It's expensive, almost impossible, for young people to get started in farming.

American Farms from Coast to Coast

Time was, an interested person could get a look at American agriculture a mile or so at a time by traveling on dirt roads by horse and buggy. Trouble was, you couldn't see enough. Today we've progressed so much that the problem is just the opposite. We see it all so rapidly, and from so far off, that farmland has become unintelligible, even to the most curious observer. Taking a jet from coast to coast when the weather is just right, which it seldom is, we can look down on checkerboard patterns of land and occasionally—over Colorado, Kansas, and Nebraska particularly—see those intriguing round circles that are actually center-pivot irrigation rigs. Mostly we see only the flight attendants or the back of the seat ahead of us, but nothing of the United States.

Time was, it was easy to drive across the country and see it: the small towns, the crops, the farmhouses, even country fairs. Today we've progressed beyond all that, too. We can roll at breathtaking speeds down superhighways, looking neither right nor left but straight ahead.

While there may be no adequate way to see all of the American farmland today, we can at least avoid the interstates and remain instead on the backroads where rubbernecking is possible. You might be surprised by what you see in a place like Kansas, for example. I once picked up a book titled simply *The Beauty*

Eastern farms are much smaller than midwestern farms. The average farm in Lancaster County, Pennsylvania, is only 84 acres, whereas the average farm in Illinois is 309 acres. Truck farms, which sell their produce at roadside stands, are more common in Pennsylvania Dutch country than in the Midwest.

Spots of Kansas. Inside the covers was nothing but blank pages. The author (or whatever the producer of a blank book should be called) had obviously never stood in a Kansas wheat field at dawn or visited Cottonwood Falls and sampled the pie in the Emma Chase Cafe. He'd never watched as a Steiger tractor dragging a 50-foot sweep turned down weeds on fallow ground to hold moisture for next year's crop. He'd never looked at the farms around towns with names like Blue Rapids, Whitewater, Alta Vista, and Belle Plaine.

Short of actually venturing forth across the countryside in your car, you can take a trip through these pages, getting an overview of the American farmland.

The East

It makes no difference where we start or end our trip, as long as we take the opportunity and the time to savor the American farmland. A good place to start is at Montauk Light, on the eastern tip of Long Island, where farmland is being paved over by urban sprawl. Although farmland is also disappearing in many other places across the country, it's most noticeable on both coasts.

Long Island, famous fifty years ago for Long Island duckling, is rapidly losing its ducks. Their production is succumbing to urban demands for land, and worry over water pollution caused by duck manure. The assistant county agent of Suffolk County, Bill Sanok, guesses that the duck population, which was 6.5 million in 1970 and is currently about 3.5 million, will dwindle to a couple of million before stabilizing.

The ducks were once raised primarily in the bay-side areas of Long Island, where the climate is milder—warmer winters, cooler summers. Alas, the bay shores have also become prime development spots, and so the ducks have gradually been moved inland, where they are raised in covered areas that provide protection from the harsher climate.

Potatoes, the other famous Long Island crop, have had an even more dramatic demise. At their peak in the 1940s, Suffolk and Nassau counties together had 60,000 acres of potatoes in production. Nassau is now out of the business entirely, and Suffolk is down to 9,000 acres.

Is anything growing on Long Island other than housing? You bet. Horticultural crops such as flowers, potted plants, and nursery stock are doing well. These crops offer a high return per acre, as the increasing population provides a voracious local market. This sequence from field crops to high-return crops is typical of farm areas undergoing urban encroachment.

Interestingly, also booming on Long Island are horse breeding and vineyards. I asked Bill Sanok about the horses. "Oh," he replied, "it's the law. New York State has public betting on horse races. Under a relatively new law, a percentage of the bettors' money goes into a pool, to be distributed to race winners bred in New York State. It's a big enough pool to stimulate breeding farms here on Long Island. Besides, the climate and soil are great for it."

Until recently, Farmingdale University on Long Island had a degree program in agriculture. But in 1987 this unique farm major was finally phased out, another victim of

Because they have little need for massive machinery and have large families to provide labor, the Amish of Pennsylvania have been able to "hold on." However, land prices in much of Lancaster County's Amish country have become so high because of urban development that the Amish are being forced to sell out and move to other, less costly areas.

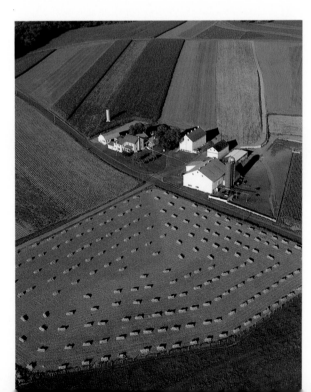

Signs on the Land

Farmers are traditionally regarded as an uncommunicative lot. But considering their buildings—frequently the major event in the landscape for many miles around—the tactiturn nature of farmers sometimes has given way to more expressive moments. This phenomenon may be due to the intervention of outsiders willing to pay for billboard space, as in the now largely disappeared Mail Pouch tobacco signs; or to some historical context, as in hex signs; or, it can just be for the fun of it. Recently there has been a trend toward what I call "crop sculpture," using crops to make designs or words that are often visible only from an airplane.

Much of farm decoration is simply innovative advertising or everyday whimsy. The hex signs of the Pennsylvania Dutch are purely decorative and may or may not have some spiritual significance. In Kansas, local farmers say that boots placed on top of Kansas fences make the fence posts last longer by preventing rainfall from penetrating the ends of the posts.

urbanization. The dairy cattle were sold off, the hogs disappeared, and farming courses were replaced by such student attractions as Food Technology, Restaurant Management, Horticulture, and Veterinary Science.

If we work our way west, bypassing the canyons of New York City and crossing over the Verrazano Bridge, and get beyond the oil storage tanks and corporate structures of New Jersey, we find farmland. There aren't many farms left in New Jersey, which has the highest concentration of population of any of our states: 986 people for each square mile. (Wyoming has only 4.9 people per square mile; Alaska has less than one person for each of its many, many square miles.) There are still lots of truck farms—overgrown vegetable patches whose products find markets in nearby cities—but high land prices are squeezing out even these farmers.

I used to know of an aggressive young farmer in Jersey who farmed several small acreages that had been bought for future development. Following a map of his rented acres, he raced his tractor along crowded highways to manage his scattered fields, dodging trucks, buses, and frantic drivers. I've often wondered what happened to him— dead, probably, either from overexertion, depression, or an encounter with a 10-ton truck on a two-lane highway.

Roll on west along the turnpikes and across the Delaware River, skirting Philadelphia, and before long you can turn off and wander the byways of Lancaster County, Pennsylvania. This is farming country, all right, even as it, too, gives way to development.

I lived in the county for thirty years, until developers surrounded me and nibbled away at our land, the fields between home and town dwindling to a solid line of suburbia. It's still a beautiful place, with rolling green fields and winding back roads, and it's wonderfully productive.

The Pennsylvania Dutch farmers are great stewards of their land; it was good soil to begin with, and they've kept it that way. I admire them, but they're doomed. Population is overwhelming them, too. The unemployment rate in Lancaster County is consistently among the lowest in the country, as new industries flocking to the county require new employees.

Fortunately, the offspring of the Pennsylvania Dutch have long since escaped in large numbers to develop other major farming areas. Mailboxes with names like Shenk, Gibbel, Schlegelmilch, and Burkholder are readily found on the backroads of almost every good farming region throughout the country.

There are almost 3,000 counties in the U.S., and in rankings for agricultural production, Lancaster County is first in production of hens and pullets, second in hog and pig production, fourth for milk cows, fifth for horses and ponies, ninth for broiler chickens, and tenth for cows and heifers. Because these are all animal crops, most can be compressed into small acreages by buying feed grown elsewhere. After all, chickens could be grown in midtown New York City, feed in and manure out, if it were economically and politically sensible to do so!

While there are still many field crops in Lancaster County, there is no field crop production that matches the rankings of animal production. This change from crops that demand big acreage to specialty crops with a high value per acre is typical for farmland on its way to being developed into cities. The

Whereas the farm areas of Long Island and New Jersey are mostly quite flat, the land in Lancaster County is more rolling. Soil conservation efforts in Lancaster County are nationally famous. Farming in strips and on the contours of the land minimizes erosion.

crops change from big-acreage demanders like corn, wheat, sorghum, and soybeans to specialty crops with a high value per acre. Wheat, for example, may bring only $150 an acre, while strawberries may bring several thousand dollars.

Sad to say, the intensive animal farming creates a difficult problem: how to get rid of the manure. It smells, it pollutes, and it is produced by the thousands of tons. One dairy cow excretes about 60 pounds of manure a day, and Lancaster County's inventory of dairy cows runs to about 90,000—that's 5.4 million pounds of wet manure to dispose of each day, plus the manure from hogs and chickens. No wonder the fishermen of Chesapeake Bay, the outlet for the Susquehanna River, which flows through Lancaster County, are screaming about pollution caused by

the concentration of animals in Lancaster County! Although environmentalists are still debating how much of the bay's pollution is caused by animal manure, how much by human sewage, and how much by industrial pollution, it is clear that the famous oyster beds have been decimated.

Odor has become one of Lancaster County's most troublesome zoning questions: everyone wants to live in the beautiful countryside, but no one is willing to tolerate hog, chicken, steer, or dairy odors or, for that matter, weedkillers and insecticides.

The pressure of development drives up the price of land so that farms that have been in families for generations tend to get smaller and smaller. Lancaster County land is regularly bringing prices of $5,000 to $10,000 an acre, while farmland in Iowa that is of equal

In the Midwest, not only are farms larger, but crops are also. Iowa and Illinois together produce more than a third of the nation's corn—the crop that brings the most total dollars. The flat farmland is immensely productive, more so than our economy can tolerate.

Farther west, water and the lack of it become more critical. The occasional drought that inconveniences farmers in the East may produce a total disaster in drier Kansas or Nebraska, where the margin of safety is narrower. This sorghum field in Kansas (right), hit by drought, is not worth harvesting. Rangeland (below) can get by with less moisture than cropland, but the pasturing of dry rangeland must be carefully controlled.

growing quality struggles to get $1,000. With land at $1,000 an acre, it's difficult to pay off a mortgage, but with land at $10,000 an acre, it's impossible to resist the temptation to sell to the developer.

Lancaster County is still a lovely reminder of things as they were, and its farms are still highly productive, but its decreasing farmland is an oasis in the middle of encroaching suburbia. Much of the East Coast is subject to the same sorts of pressures as Lancaster County. We'll see the loss of farmland everywhere the population is increasing. In Florida, for example, where so many "Northerners" have settled, the state's population has exploded from 3 million people in 1950 to 10 million today.

The Midwest

The *real* American farmland lies to the west. Let's move on to Illinois, Iowa, Kansas, Nebraska, and then on across the country to Texas and California. In Indiana and Illinois, you can see that the farms have more acreage. The tools of farming get bigger, more expensive, and more efficient, and the crops become those blockbusters we are overproducing, stockpiling, and worrying about: corn and soybeans mostly, spare land in wheat, a little sorghum, some hay.

Drive down south of Decatur, Illinois, and you'll find no better farmland in the country: rich black soil, flat land, great productivity. No animals here, just corn and soybeans. I talked with the county agent there, who said: "Below Decatur there's hardly a farm animal bigger than a dog—oh, maybe a 4-H horse here and there. Animals interfere with leavin' the farm in the winter." CS&F farmers, they

33

Forage and corn are two of the blockbuster crops; wheat and soybeans are the others. Here alfalfa and corn are grown side by side in Iowa.

are called: corn, soybeans, and Florida. With no animals to care for, they can go to Florida in the cold weather. The farm depression has cut back on trips to Florida, perhaps, but there are still lots of condominiums that shelter Illinois farmers for the winter. Why not?

Farms are bigger here than in the Pennsylvania Dutch country. The average farm in Illinois is 309 acres, whereas the average in Lancaster County is only 84. In Illinois we're not only into the heartland of America but into the heart of that ogre, the farm problem:

overproduction, low prices, and government supports.

Farmers in Illinois and Iowa josh each other about which state is more productive, but they actually grow almost equal amounts of the major crops. Both states are immensely productive. Together they raise 35 percent of the total U.S. corn crop. Their soybean crops almost match as well, accounting for 30 percent of the U.S. total.

When it comes to animals, differences arise, because Iowa is an animal state, while

34

Soybeans and corn are the two major midwestern crops. Herbicides to control tall weeds in soybeans can be applied with a wick applicator that only releases herbicide when it brushes against a tall weed, leaving the shorter soybean plants untouched. This method saves herbicide and puts fewer chemicals on the land.

Illinois is much less so. In 1984 Iowa had more hogs on farms than any other state: 14.2 million (compared to only 3 million people). On the other hand, Illinois had only 5.4 million hogs.

As transportation has improved, and farms have grown in size and purchasing power but dwindled in numbers, the status of small towns has shifted. Many of the small towns in the big crop areas of Illinois, Iowa, Kansas, and Nebraska, as well as other midwestern states, just haven't made it. The small town seen in TV commercials, with white houses and broad lawns, kids on bicycles delivering newspapers, and a brass band playing in the town square on the Fourth of July isn't gone, thank goodness, but there are fewer of them than there were forty years ago, and many of the younger inhabitants have moved to the city.

The Plains States

So far, as we've been moving west, the cropland across the country has been much the same. After all, land in both Iowa and the Pennsylvania Dutch country can raise the same crops and animals, primarily because there is little difference in rainfall. Go

The "Old West" still exists. Cowboys ride horses, and cattle graze open range during the summer and must be fed during the winter. Ranching is probably the least changed, at least on the surface, of any area of agriculture.

south and the rainfall soars: the lower Mississippi states—Louisiana, Mississippi, eastern Arkansas—all get more rain than they want, or at least too much rain at the wrong time. Mississippi averages almost 53 inches of rainfall, and Louisiana gets just shy of 60 inches, whereas the East and Midwest get 30 to 50 inches. We'll look at the South in greater depth when we discuss their cotton, rice, and tobacco crops.

Drive west, however, to Kansas and Nebraska, and you'll see major agricultural differences. West of Illinois and Iowa the rainfall diminishes, and the crops change with it. Look at Nebraska, for example, and study the precipitation changes occurring within the state. Omaha, along the Missouri River on the eastern boundary, averages 30 inches. But Scottsbluff, 400 miles west in the Nebraska panhandle, has an annual average of only 14.6 inches—less than half that of Omaha. To the south, Kansas shows the same shift. Topeka gets nearly 30 inches of rainfall, while the towns along the Kansas-Colorado border get only 16 inches.

This is the nitty gritty of it: water availability determines what we grow. The rule of thumb is that below 20 inches of annual precipitation there are few crops that can grow without irrigation. Wheat is the foremost crop that can grow under these conditions; the other blockbuster crops just don't make it without supplementary water. Sorghum does better with less water than soybeans or corn; cotton can make a crop, but not a hefty one, with 16 to 20 inches of rain. Wheat, the dry farmer's friend, can produce a crop with as little as 14 inches if it comes at the right time.

Traveling west from the Missouri River on Nebraska's eastern boundary, for the first 50 miles we see corn and soybeans, which are raised mostly without irrigation. Just west of the midway point in the state, the rainfall level drops to below 20 inches, and irrigation gradually becomes more routine as we move west. By the time we get to Scottsbluff in the western panhandle, the average moisture is down to 15 inches, and only wheat or barley can be grown without irrigation.

Of Nebraska's 21 million acres of cropland, some 6 million are irrigated. Half of this irrigation is center-pivot irrigation, those green circles you may have seen from an airplane seat while flying over the Midwest.

Western Nebraska sits atop the north end of the gigantic Ogallala aquifer, and Nebraska's irrigation wells reach down into this huge pool of underground water that has accumulated over many centuries. In the next chapter we'll see more of what it means to be over the Ogallala, for it extends south through western Texas, making agriculture possible in the fields above it.

Irrigation not only makes crop production possible but sometimes can make it superb. Irrigated crops usually achieve higher yields than naturally watered crops—higher yields, but not necessarily higher profits. In 1984 Iowa averaged 108 bushels of corn to the acre; Colorado, where corn is irrigated, averaged 141 bushels. This isn't without a price, of course. Rain is free, but irrigation in 1984 cost Colorado farmers more than $27 million.

As we move out of the irrigated areas along the Platte River in central Nebraska and into the Sand Hills country to the north, we come upon the legendary grazing land where ranches are big enough, and far enough from town, to have their own airstrips, and where cattle can wander over thousands of acres of rough, humped land. The land is excellent for grazing, but it's dry country that necessitates large acreage per cow. Although the ranches here are of necessity gigantic, they are not necessarily more profitable than smaller spreads with better rainfall.

In western Nebraska we come into wheat country, the area that takes advantage of wheat's ability to survive with little water. My favorite wheat area in the entire country is west of Scottsbluff, partially in Nebraska, partially in eastern Wyoming. For miles and miles the vast high tableland is blanketed with alternate strips of fallow ground and burgeoning wheat. Like much of the American wheat country, rainfall is so scarce here that the land is only planted every second year. In its off year the land is left fallow, unplanted, and the weeds are chopped and turned under to conserve moisture for the crop year. When a farmer says he farms a thousand acres here, he usually means he has a crop in 500 acres and has 500 acres fallow.

Much of this marvelous wheat country is surrounded by towering buttes. Farm lanes might be a mile long, leading off a dirt road that can be pure slick during a rainstorm. Eagles are likely to be found soaring along the edge of the buttes. From an airplane in midsummer, this high country seems to go on endlessly in narrow alternating strips of tan wheat and brown, dusty soil, broken by an occasional butte or gully and a rare farmhouse with outbuildings surrounded by a few hard-to-come-by trees in a shelterbelt. This is

Compared to the smaller farms of the East, the farms of the Midwest and West seem to stretch endlessly. Irrigation equipment becomes commonplace, rangeland opens up, and cattle feeding seems to be at every crossroad. The need for fertilizers and other chemicals, feed, and equipment is gigantic. Far more people are involved in supplying farmers or in processing farmers' crops than in farming itself.

lean country, but when moisture is right, it produces a lot of wheat, and it's a beautiful sight.

To the south, the huge state of Texas has much of its best production centered in the western panhandle, an area that gets just shy of 20 inches of annual precipitation. Like western Nebraska, the panhandle lies on top of the Ogallala aquifer, which is gradually being sucked dry by the demands of the panhandle's irrigation pumps.

There's lots of experience here with dry-land farming. During the Dust Bowl of the Depression years, farmers watched their fields being blown away. Although much of this land went out of production in subsequent years, in the late 1970s the plow got to a lot of it again. This was the period when we thought that surpluses were gone forever,

that exports of farm products had no end, and when the Secretary of Agriculture said to "plant fencerow to fencerow."

Much of the panhandle is devoted to crops for cattle feeding. At any time the area around Hereford, Texas, has "on feed" a million head of cattle. An Easterner who had business in Hereford disembarked from his airplane flight, sniffed the air, and asked his host what the ghastly smell was. "That, sir," the Texan replied, "is the smell of money."

The Rocky Mountain States

When we reach the Rocky Mountain states, the decreasing precipitation level and changing landscape become obvious. There are a

As you move west beyond the 20-inch rainfall line, which stretches through the middle of Nebraska down to the panhandle of Texas, water becomes more scarce and the farming becomes less intense. Dry-land farming is more risky as plowing opens land to wind and water erosion (far left). Much of this land is left in natural grasses, useful for grazing (above).

The most successful nonirrigated crop is wheat. Usually grown in "fallow-farmed" strips (top), nonirrigated wheat is often grown on dry ground adjacent to irrigated crops (bottom).

few level areas with irrigation, areas like Weld County, north of Denver in Colorado, and large acreages of dry-land wheat in the rolling foothills or occasionally in mountain valleys. Mostly, however, the agriculture of the Rockies is based on ranching.

The outstanding difference here is that, as the land becomes drier and the topography steeper, more of it belongs to the federal government. I live on the eastern slope of the Continental Divide in central Colorado, and more than 70 percent of the land in our county is federally owned. Originally, of course, it all belonged to the federal government, then from 1781 to 1982 homesteaders were either given or were sold 287.5 million acres—half a million square miles—which is an area almost twice the size of Texas. As an incentive to build, railroads were given 94.4

sorghum, and soybeans. They latch onto crops that start out with high markups, and because of the state's immense population (it has the largest population of any state, having passed New York in the 1970 census), the short shipping distances from California farms to local markets give California farmers an advantage in products such as milk and eggs. Because of its unique climatic variances, California has year-round production of many crops, particularly vegetables.

The saga of how California transformed from desert to agricultural producer is mostly a story of politics and pressure during the era when big dams were being built in unbelievable numbers and sizes. Without the dams and their water storage, California's cities could never have grown, and its agriculture would have been negligible, for both depend on water sources that aren't naturally available in the arid California climate. Many of the big farming areas in California get less than 8 inches of rain a year.

California is reputed to be the home of corporate farming, but in fact this is not an accurate assessment. According to the 1982 *Census of Agriculture*, there were 82,463 farms in California, but less than 1 percent of these was owned by other than a family. In acreage, however, the figures were different. Of the total 32 million acres in farmland in California, the 809 farms owned by other-than-family corporations comprised almost 5 percent. This is still not overwhelming and

certainly does not support the frequently held idea that California farming is entirely in the hands of big corporations.

It is certainly true there are a handful of giant corporate farm enterprises in California. Newhall Land and Farming is a good example, although not the largest, owning 123,200 acres of land on eight ranches in California. While this corporation is primarily interested in real estate development, it raises grapes, rice, tomatoes, alfalfa, sugar beets, beans, cotton, and cattle. In 1985 its agricultural sales came to $40 million, while its residential, commercial, and industrial sales topped $100 million.

California dairies are usually pictured as huge factories, with giant milking parlors used round the clock, milking 2,000 cows or more. A number of these dairies certainly exist, marvels of efficiency, but the average dairy farm has only 204 cows. Even at that, California's dairies tend to be much larger than Wisconsin's, which average only 42 cows per farm, an interesting figure when you consider that Wisconsin is by far the largest producer of milk, with California second.

California brings us to the end of our westward trip. Before we can understand the process and problems of farming, we need to take a closer look at the immense importance of water, how different areas get it and use it. Then we'll go on to look at specific crops, how and where they are grown, and how they are marketed.

Water

If you farm in the East, you don't give much thought to water. Rain is plentiful. Sure, there are occasional worries about droughts such as the dry time in Georgia in 1986, but the droughts are ordinarily short and recovery is quick. If you farm in the West, however, water is at the top of your list of concerns. Without it farming is, at best, impoverished; at worst, impossible. Nature has been niggardly with precipitation in much of the West, adequate to sumptuous in the East.

The line of 20-inch rainfall, which runs vertically through the middle of Nebraska, down through the Texas panhandle, is usually considered the magic dividing line, east of which nature provides, west of which man takes responsibility for moisture. Of course, people don't actually provide water in the West; they simply move to more useful locations the water that nature provides.

On a national scale, how important is supplemental water to farming? That depends on the crop. In an average year we produce about 7.5 billion bushels of corn, which is our largest single crop in terms of dollars. More than a billion bushels of this total is produced on irrigated ground. If we had no irrigation, we would pretty well do away with our surplus of corn, which might seem like a great idea—but don't suggest it to the irrigation farmer.

Other crops, because of where they are grown, are more dependent on water, although the dollar value of minor crops hardly matches that of corn. If irrigation were not possible in California, Arizona, and southern Texas, specialty crops like avocados, kiwifruit, artichokes, and even oranges would be confined to the diminishing subtropical acreage of Florida. Our diet would be seriously affected. Without irrigation we would hardly starve, thanks to the surpluses we are currently producing, but we would have far fewer choices of foods at vastly higher prices.

About 83 million acre feet of water are used for irrigation each year. The water used annually in irrigating our crops thus comes to more than a hundred thousand gallons of water for every person in the country. Of this, some 44 percent is used by the three top irrigation states: California, Nebraska, and Texas. California alone uses 24 million acre feet of the total.

What are the sources of supplemental water? If it is pumped from under the ground where nature has stored it, we call it groundwater; if it is moved on the surface from runoff created by rain and snow, through dammed reservoirs, streams and rivers, canals, and ditches, we call it surface water. About 44 percent of the nation's irrigation water comes from groundwater sources: wells. Fifty-six percent is from surface water that has been moved anywhere from a few hundred feet to a thousand miles.

The biggest difference between groundwater and surface water is that, given good weather conditions, surface water is replenished each year by rainfall and especially by mountain snowfall. Groundwater, on the other hand, has accumulated over thousands or perhaps millions of years, so its annual replenishment rate is very low. Renewal depends on the soaking in of rain and the absorbing of water from streams that flow across the surface. Renewal rates vary immensely, but there are few serious irrigation areas where the renewal matches withdrawal. That's one of our major agricultural problems.

46

Groundwater

Until the centrifugal pump was invented about seventy years ago, groundwater irrigation was dependent on primitive techniques to lift the water. A new era began when electricity was harnessed to the centrifugal pump. There are now an estimated 317,000 irrigation wells in the United States.

The cost of drilling a well to tap groundwater has been a private cost borne by the landowner who wants the water. By contrast, for the most part the cost of harnessing surface water has been a public cost borne by the taxpayer in the often unfounded belief that the money will be repaid by the farmer through usage fees. The repayment record isn't a very good one.

The political porkbarrel of dam building for surface water control has until recently been almost unstoppable. What politician could vote against water for his district? What form of political suicide could be easier? Times have changed, however, and the era of big dam construction has likely disappeared.

Well drilling is increasingly controlled as states begin to inventory and be concerned about their groundwater. In some states a hard-to-obtain well permit is necessary, but in others it is still relatively easy to drill into the aquifers and pump out major quantities of water.

The Ogallala

The story of groundwater can be largely understood by taking a look at the Ogallala aquifer. An aquifer is a water-bearing layer of porous rock, sand, or gravel. More than 2 million years ago the Ogallala was a seabed

Irrigation water in the West is often moved hundreds of miles through a complicated series of dams, rivers, canals, and ditches. This surface irrigation water is mostly snow melt from the mountains, which has turned former deserts into superb farmland.

extending from western Nebraska and eastern Wyoming, south through western Kansas and eastern Colorado, western Oklahoma, western Texas, and eastern New Mexico, almost to the Mexican border. The aquifer has been absorbing moisture very slowly over millions of years. Groundwater scientists estimate it currently holds about 3.2 billion acre feet of water, and the entire United States uses only a little more than 80 million acre feet a year. Nevertheless, the level of the water is dropping.

Although it can't be measured very accurately, the thinking is that the Ogallala can recharge itself only about half an inch to 2 inches a year. The thickness of the water-bearing aquifer varies greatly, from as little as 50 feet on the southern edge in Texas to 1,200 feet in Nebraska. If the ability of the aquifer to build its water reserves is miniscule, the ability of thousands of wells with centrifugal pumps to lower the water level is tremendous. Although it is estimated that only 5 percent of the aquifer has been tapped, water levels in many areas are dropping seriously, especially in Texas.

As the water levels drop, wells have to be deepened to reach the water, and pumps have to work harder to bring the water up. This has already brought many thousands of acres in the aquifer to the point where using water to farm the existing major crops of the area— corn, sorghum, cotton, and hay—is of questionable profitability.

In Texas the problem is much worse because the cost of the natural gas that fuels most of the pumps in western Texas went from thirty-seven cents a thousand cubic feet in 1973 to four dollars a thousand in 1985—

an increase of 1,100 percent. For this reason many west Texas farmers are looking at the possibilities offered by new technology for minimizing groundwater use or doing without it entirely.

Much of the Ogallala area in western Texas receives almost enough rainfall to farm without irrigation; the 20-inch rainfall line comes down through the eastern edge. Farming there without supplemental water is risky, but cheap. Is it better to produce two bales of cotton to the acre with a relatively sure crop but high water costs, or one bale of cotton with some risk of no crop at all but no water costs? These are questions that the coffee drinkers in the roadside cafes around Plainview and Lubbock in Texas are asking one another frequently these days, and there are plenty of changes already in the works.

Farming a lot of this acreage without irrigation is legal, but it's dangerous, although not as dangerous as it was in the Dust Bowl days of the 1930s. There's a lot more known about how to handle the dry soil, and more crops have been adapted to do better in dry conditions. For the American not dependent on farming for a living, it might seem that the production from this risky land is hardly needed when we are already spending billions of dollars to convince farmers to grow less. But for the farmer whose family has farmed a piece of ground for a hundred years and whose livelihood may disappear, the question is less simple.

The federal government isn't sitting idly by. In the Food Security Act of 1985, the government stepped in to make it financially attractive to the owner not to farm the worst of the highly erodible land but to rent it to the

California uses the most irrigation water—almost 30 percent of the total in the United States—and is almost totally dependent on irrigation. Many states such as Kansas, Nebraska, Colorado, and Texas, while they do not need to depend entirely on irrigation, have greatly enhanced their growing capabilities by the use of supplemental water.

government under the Conservation Reserve Program (CRP). The goal is to have 40 million acres in the United States in the reserve by 1990. How much of this will be in the Ogallala aquifer area of Texas isn't yet known.

The ingenuity of farmers in becoming more productive on less land is legendary. We're currently producing roughly two and a half times the farm products we did in 1930, using almost the same amount of land. The harbingers of doom who have been shouting that the farmer's days of increasing productivity are over haven't visited any agricultural research stations and certainly haven't walked the furrows with the best of our farmers.

Even before the Conservation Reserve Program, the decline in groundwater levels and the increase in pumping costs had already lowered the number of irrigated acres in the Ogalalla area of the Texas panhandle. In 1977, 6.4 million acres were being irrigated; by 1984 the number had dropped to 4.5 million. Land prices, of course, were affected. Irrigated ground is simply worth more than dry ground; it grows better crops, and farmers are willing to pay more for it. Increase water costs or do away with water, and land values will drop.

Is there a crisis? Lots of journalists and not a few politicians argue that there is, but the answer really depends on who you are. The lower production from farming without irrigation or from putting land into the reserve program will help, however little, to keep down the nation's surpluses and solve our farm problems. If I were the farmer who owned the land that was going dry, however, I would shout *Crisis!* at the top of my voice.

One of the great realizations about changes in agriculture is that they seldom happen overnight. Market and cost considerations are likely to cause gradual changes—evolution not revolution. The deepening of wells in the Texas Ogallala area has been going on for many years; land prices were affected long before anyone was changing farming techniques. Change, fortunately, develops a slow rhythm of its own.

Surface Water

If groundwater sources account for almost half the irrigation water in the United States, what about the rest? In Texas 73 percent of the acres of irrigated land have wells as the only irrigation source; on the northern end of the Ogallala, in Nebraska, the figures are almost the same. But in California the opposite is true; only 21 percent of the irrigated acres have wells as an only source. California depends on surface water, much of it brought from outside the state.

Remember that of all the irrigation water applied from all sources in the United States, California uses almost 30 percent. With the exception of a few areas in the northern part of the state, California is totally dependent on irrigation water for its agricultural living, and its agricultural living is a giant part of the state's economy.

Logically, but somewhat surprisingly, although California uses almost one-third of the irrigation water in the United States, the acreage it irrigates is far less than that of second-ranking Nebraska. This seeming anomaly is explained by the rainfall map. Nebraska and Texas get far more rain, need far less irrigation per acre. California puts on

Slightly more than half of the irrigation water comes from surface water supplies, which often have to be moved hundreds of miles. In some places, however, such as this valley in western Colorado, the water is put to use almost directly from the snow melt in the surrounding mountains.

Types of Irrigation

It is difficult to know whether getting water to farms in dry country is more important than knowing how to use it once it is there. Most surface water comes to dry land through government-financed projects; ground water comes via privately paid-for pumps. Either surface or ground water can be used as the supply for a varied set of water application methods. Early irrigation depended mostly on gravity feed right into the field, but the development of power-operated pumps made new irrigation techniques possible. Now emphasis is on more effective utilization of increasingly scarce water.

Gravity flow irrigation is being supplanted by center-pivot systems that have less water loss, but cost more. Drip systems that run water through small hoses at low pressure, with specific application points, conserve water better than any other system, but installation costs are high also.

California, the biggest irrigator, has about 70 percent of its irrigation as gravity flow, 23 percent in center-pivot systems, and about 6 percent in drip irrigation. Nebraska, the second largest irrigator, has no appreciable amount of drip irrigation, but delivers water to more than half its irrigated acreage from center-pivot systems.

Facing page: Siphon tubes move water from ditches into fields, either to flood the fields for crops such as grains (top left) or to irrigate rows for crops such as corn or vegetables (lower left). Surface water is moved to the fields through canals (top right) or is pumped from groundwater sources (lower right).

Drip irrigation, which is applied in small quantities directly to crops through plastic tubes (top right and left), is used mostly on fruit and vegetable crops. It saves water but is expensive to install. Center-pivot irrigation (bottom) is extensively used, requires minimum labor, can be used without extensive field leveling, and allows great control over the amounts of water put onto a crop. It looks beautiful from the air, too!

an average of 2.8 acre feet of water per acre, Nebraska only 1 acre foot. Bakersfield, in the heart of much of California's agriculture, gets less than 6 inches of annual precipitation. Without irrigation it qualifies as desert. The middle of Nebraska gets 20 inches. Less rain, more irrigation—logical.

Where does California's water come from, and how does it get there? Much of it can only secondarily be called California's water; it originates somewhere else. For the most part the water comes from mountain ranges that catch winter snows: the Sierras in eastern California and the mighty Rocky Mountains. How it gets to the farming areas of California is a saga of need and greed, of engineering and determination and above all, of politics. The mass movement of water through irrigation projects was an idea that arose in the 1920s and 1930s and is nowhere better exemplified than in the Colorado River.

The Colorado River System

The Colorado originates in the high country of northwestern Colorado or in southern Wyoming, depending on which tributary you accept, and ends 1,450 miles later in the Gulf of California. At least what remains of the river empties into the Gulf of California, for there isn't much water left by then. It's been used, abused, diminished, and salted. The drainage basin area of the Colorado includes seven states: Colorado, Wyoming, Utah, New Mexico, Nevada, Arizona, and California. Its final 80 miles before flowing into the Gulf of California are in Mexico.

It's not a giant among rivers, not compared with the Mississippi or dozens of other American rivers, but the Colorado is a river that does things in a big way. Before it was harnessed, it was a wild one among rivers. Now that it has been harnessed, it's resentful.

When accidents happen on this river drainage, they are likely to be of giant proportions. In 1905, just below the California-Mexico border, a break in the bank diverted the entire river from its course through Mexico and channeled it back into southern California, dumping it into a huge sink more than 200 feet below sea level.

Hard to believe, the Salton Sink became the Salton Sea, which gradually became a salty lake. As the water evaporated, it left the salt behind (the lake currently has about the same salt content as the ocean). The Salton Sea became more than 10 miles wide, 50 miles long, and about 70 feet deep. The break eventually got repaired, the river went back to the Gulf of California, and irrigation drainage waters from the neighboring Imperial and Coachella valleys have flowed into and stabilized the water level of the Salton Sea, which has become a huge recreation area. The break in the river might have inundated the entire Imperial Valley had it not been controlled. Surprisingly, it was brought under control by a railroad company protecting its right-of-way.

The Colorado's waters are contained by a large number of dams. The Hoover Dam, which forms Lake Mead, was an architectural wonder when it was completed in 1936 and remains a huge tourist attraction. Since then a great many storage dams have been built along the Colorado, engulfing about everything except the Grand Canyon, and ideas for a dam that would do away with the canyon's white water pop up frequently, only

Dams like the Morrow Point Dam on the Gunnison River in Colorado create extensive reservoirs and are the backbone of the Colorado River water-storage system.

to be beaten down by outraged environmentalists.

It's likely that for the foreseeable future the building of more huge dams on the Colorado, or elsewhere for that matter, will run into tough opposition. With agricultural surpluses almost burying us, it's difficult to justify dam projects that create new agricultural acreage at a cost of $3,000 to $10,000 an acre, especially at a time when excellent midwestern farmland, watered directly by nature, is bringing less than one thousand dollars.

The Colorado water, about five years prior to the building of the Hoover Dam, was apportioned among the seven states through which it flows. Depending on whose viewpoint you take, the upper river states got more water than they can use, the lower river states got less water than they need. In the interim since the building of the dam, the lower states—Arizona, Nevada, and California—have soared in population, partly because of the availability of Colorado River water. The upper states—Utah, Colorado, New Mexico, and Wyoming—have grown, but not to match the southern states; in fact, the areas of the northern states actually drained by the Colorado have lagged in growth.

While the states have been scrambling for more of the Colorado's water, Mexico has been a not-so-quiet complainant, for in effect it gets what's left. In 1945 Mexico got a guaranteed reserve, which is theoretically 1.5 million acre feet. Not only does it have trouble getting the water, but the quality makes it almost useless, for the Colorado has a higher than normal salt content everywhere except the headwaters, and the salt content increases as the river flows. Part of the saltiness is natural, part the result of irrigation that diverts the water through salt-laden soil and back into the river. The best thinking is that about half the salt is natural, half induced by irrigation practices. In the works is a giant desalination plant, largest in the world, to benefit the quality of the water going to Mexico.

Water, or the lack of it, will be a major economic and political problem for the entire West in the foreseeable future. We cannot satisfy the increasing demands from growing urban populations and at the same time meet agricultural needs. Because of surplus farm production, it seems likely that farming will have to give up part of its water. It can afford to, since it is estimated that at present as much as 90 percent of the water is used by agriculture. Doubling urban supplies would only mean decreasing farming's share by 10 percent, but don't look for farmers to give it up without a struggle. And don't be surprised if both sides try for more giant dam projects.

The Blockbuster Crops

There are some crops that are grown in such greater quantities than any others that I call them blockbuster crops: wheat, corn, soybeans, and hay. Most of us give no thought to the difference in the importance of crops, but to understand American farming, we should. The blockbuster crops are often the ones we have most in surplus, and it is important that we see them in perspective with the rest of what grows in the United States.

The blockbuster crops occupy almost equal giant acreage, with corn occupying the most space and returning the most dollars. Although the acreage of some is decreasing thanks to federal programs, the average for each of these crops is more than 60 million acres, an area ten times the size of Vermont or one-third the size of Texas. In total, the four crops occupy a space larger than all of Texas.

Although they all occupy almost the same acreage, the value of the crops varies greatly because of yield differences. In 1984, for example, corn brought more than $20 billion, wheat less than $9 billion. Corn produced an average of 107 bushels to the acre throughout the country, but some fields did twice that well. Wheat produced slightly less than 40 bushels per acre, although irrigated wheat in some areas might have done 100 bushels, and some marginal wheat in other areas undoubtedly was just plowed under, abandoned.

One of the great truths of our entire crop production is that so many of the major crops are, if not interchangeable, at least substitutable and thus highly competitive. Soybean oil and corn oil are both usable in margarine, for example—not to mention the half-dozen other oils. And if the price of wheat gets too low in relation to the price of corn, wheat begins to be fed to animals.

The blockbuster crops are not necessarily interchangeable in relation to land use. Wheat can in most cases grow where corn grows, but it usually doesn't because the revenue per acre from corn is so much higher than from wheat. Wheat is therefore ordinarily reserved for areas where corn can't be grown, usually because there's inadequate moisture. Hay, like wheat, needs less rainfall than most crops. It averages about 2.5 tons to the acre in the United States, but 7 to 10 tons is not uncommon in areas that have plenty of moisture and a long growing season. A Nevada or Colorado rancher without irrigation may cut hay once; a Nebraska farmer with irrigation may make three or four cuttings. The goal isn't necessarily the greatest pro-

The four blockbuster crops are soybeans, corn, forage, and wheat. Each occupies about 60 million acres, but the crops vary greatly in total value, with corn bringing the most money.

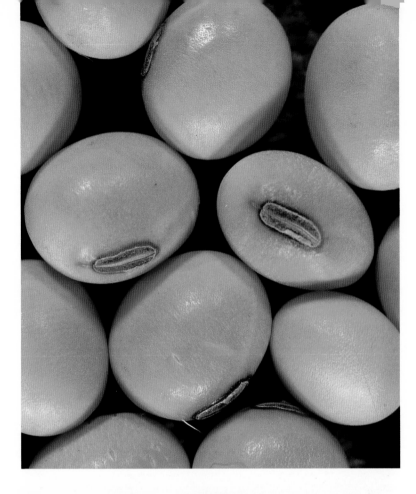

duction, but the greatest production with the least cost.

One of the marvelous features of American farming is that the right crop seems to evolve for the right ground. In southeastern Wyoming, where the rainfall is so skimpy that crops can be produced only every other year, wheat has been found to be the only possible crop, and farmers happily add to the 60 million acres of wheat already in production. This is what the "farm problem" is all about—overproduction—and it's the blockbuster crops that are mentioned most often when the farm problem is discussed.

Although the blockbuster crops seem well established, it's impossible to believe that any crop can't be toppled, replaced by something newer. Proof of this is the rise of the soybean, which was hardly known in the United States before 1920. There wasn't even an association for soybeans until 1925—and in agriculture a crop can't be anything until it has a promotional organization—then the crop took off. During World War II production doubled. By 1968 it reached a billion bushels, which by 1979 had doubled to 2 billion bushels. Now production has leveled off, due mostly to international competition.

Will some presently almost unknown crop be a new blockbuster in another fifty years? We've gotten better at knowing about unusual crops, but there is still room for something presently unknown to zoom to the top level—and there are lots of farmers aching to try something new, exciting, and at least momentarily profitable.

Wheat

If you turn north from I–70 just east of Denver and drive through the village of Bennett, then on up Route 79, and if you're there at the right time of year, the vast, open, treeless plain stretched before you seems alive. Alive with the biggest tractors you can imagine: bright red Versatiles, behemoths originally imported from Canada—the Kalcevic family has more than a dozen of them—and light green Steigers from a factory in North Dakota—the Lewton family has seven of them in a field at once. Dust plumes from the tractors rise across the plains like a series of Indian smoke signals.

This is wheat country: tough, raw, but if everything goes right, very productive. That's a big "if," but the farmers find the payoff worth the gamble. And there is wheat country like this not only in Colorado, but in many other dry places as well: the Palouse in eastern Washington, where the hills are so steep that tractors must have cleats instead of wheels to run the tillage equipment, or the dry country of the Texas panhandle, places like Deaf Smith County, where the soil blows into the next county if it's left without a crop to cover it.

Wheat can grow in richer country than this, of course, and at least a few acres are grown in every state, but what's important here is that it can grow in wild, arid country that, without supplemental water, is good for little else. Only corn takes up more acreage than wheat. Wheat covers more than a hundred thousand square miles of cropland, an area twice as big as all the New England states. Although wheat is grown to some extent in every state, of our total production of

2.5 billion bushels, Kansas is the leading grower, with more than 400 million bushels, followed by North Dakota, then Oklahoma.

Wheat reaches back into the mists of prehistory. There is evidence that strains of cultivated wheat existed in the early Stone Age; a crude sort of milling equipment believed to date back 75,000 years is known. The earliest actual remains of the grain itself date from about 5000 B.C.

Products made from wheat provide more of the basic diet for the world than those from any other crop. The nearest competitor is rice, which is geographically limited because it is the opposite of wheat, demanding large amounts of water for its cultivation.

Curtis Lewton, whose Steigers help churn the ground on the flat prairie east of Denver, is a thirtyish man, solidly built, whose skin is tanned from the endless wind and sun he's grown up with. Curtis has a soft but clear way of talking, and he chooses his words carefully, thinking out what he wants to say before he says it. There isn't much excitement in his voice, but it carries a chuckle with it. Curtis's grandfather came to Colorado from Kansas in 1919 and bought land from the railroad.

"I asked him many times why he came here," Curtis smiles. "I got a different answer most times, but the most consistent one was that he was tired of trees and rocks. He chose well, we don't have either here."

Curtis's father, Glenn, and his late uncle, Harold, farmed the land and expanded the original section (640 acres) that Curtis's grandfather bought from the railroad. Today the land is owned by Glenn, Harold's estate, Curtis, and Harold's son, Wayne. Each farms

some of the acreage individually, and they jointly farm the land they own as a family corporation. All told, the land they farm comes to something more than 13,000 acres—or more than 20 square miles—including some ground that they lease. They are a family farm, but a big one.

About half of this acreage is "fallow farmed," left unplanted every other year to conserve moisture. This is a typical practice in wheat country, where moisture drops much below 20 inches annually.

"The statistics say we get about 15 inches of rain," Curtis explains, "but some years we have more, some less, usually less. We've been down as low as 10 inches. In fact, one year back in, I guess it was in the thirties, my Dad remembers we went 135 days here in the summer and didn't have as much as a sprinkle, and we also didn't have a crop. Then, in 1955 I think it was, Uncle Harold took one combine to the fields and cut all of the wheat in a single combine tank and drove back home. That was the crop for the year."

Wheat's greatest asset is that it does well with the heavy rainfall of the East and Midwest but can get by with as little as 14 inches a year if the rain and snow come at the right times. Wheat may not be perfect, but it sure comes close. Feed irrigation water to it just right, and use the newest varieties, and you can top 100 bushels of grain to the acre. Skimp by with the moisture that falls naturally in the 14- to 20-inch rainfall belt—eastern Colorado, western Kansas—and you can come up with 25 to 50 bushels; some years more, some years less, a few years none.

In the United States about two hundred varieties of wheat are grown commercially, but

There are about two hundred
varieties of wheat grown
commercially in the United
States. This is a "bearded"
variety of hard red winter
wheat.

the Lewtons usually plant only two or three varieties. "We planted Hawk and Vona the past few years," Curtis explains. "I listen around and look at the college's test plots, and if I see something that's doin' better than what we have, I try some."

For growing and marketing purposes, the United States has settled on grouping wheat into five classes according to the kind of flour produced and the time of year the wheat is planted. The class is determined by the hardness of the kernel, the color of the kernel, and the planting time. (The color of the kernel, incidentally, has no effect on the quality of the wheat or its use in flour.) The five categories are hard red winter, hard red spring, soft red winter, soft white winter, and durum. The winter wheats account for 80 percent of the total production, spring wheat about 17 percent, and durum accounts for the final 3 percent.

Winter wheat is something of a misnomer, for it is planted in late summer or early fall and "winters over," to be harvested the following summer. Spring wheat, however, is planted in the spring and harvested late in the summer of the same year. Spring wheat is grown in the colder climates of the states that border Canada, where young wheat plants would freeze out if they had to winter over.

In states where winters are not too severe, winter wheat offers a fortunate advantage to farmers: it sprouts in the fall and develops leaves, which can be eaten by cattle throughout the winter and on through the spring. The root system then sends up new growth and the wheat matures in spite of its having been eaten back to the ground—so the farmer actually gets almost two crops for the cost of one.

In many parts of the southern end of the wheat belt, cattle are routinely fed on wheat pasture for four to six months. Recently I was down in the panhandle of Texas, east of the cattle-feeding center of Hereford, when a bunch of heifers came off pasture and were loaded into potbelly trucks to be shipped to a Kansas feedlot, 500 miles east, for their "finish" feed.

"I put them in here on the wheat on the fifth of November at an average of 510 pounds," the owner told me, "and they're coming off pasture today, the twenty-third of March, at 800 pounds; that's more than 2 pounds a day gain from the wheat pasture. That's pretty damned good! They'll grow faster in the feedlot, but that's more expensive feed."

Hard red winter and hard red spring wheats, both high in protein, are used for bread flour. The soft wheats—soft red winter and soft white winter—provide the flour for cakes, pastries, crackers, snacks, and breakfast foods. The West and the East Coasts are the most prolific sources of these. Much of the soft wheat produced in Washington and Oregon is shipped directly to West Coast ports for export to Japan and other Asian countries. (Japan, incidentally, has often been the largest importer of American wheat, surpassing the Soviet Union, whose imports make greater headlines.)

Durum is the hardest of the wheats. It is also high in protein and provides the semolina and durum flour for macaroni, spaghetti, and noodles. Durum is grown almost entirely in the northern tier of states, along with hard red spring wheat.

In warmer states, winter wheat serves as a double crop. Planted in the fall, it grows until freezing weather. The leaves, usually a few inches to a foot tall when growth is stopped by freezing, can be grazed by cattle. In late spring, when the cattle are removed, the grain head grows relatively unharmed and produces a grain crop.

Wheat planting is usually done with a "drill" that opens a slit in the ground, drops seeds into the slit, then closes the slit. Seeds can be drilled from 1 to 6 inches deep. Winter wheat is planted in the fall, and the wheat matures the following summer. Spring wheat, mostly limited to the states close to the Canadian border, is planted in the spring and matures the same summer.

Planting

Because wheat isn't a fussy crop, the ground can be prepared for planting in a variety of ways. Although wheat is usually planted in a soil that's been finely worked, there's a move nowadays toward not tilling the soil at all after the last crop. Instead, herbicides are sprayed on the field, then the new wheat seeds are knifed right into the ground, in the stubble remaining from the earlier crop. The merits of this "conservation" planting, which conserves the soil from wind erosion because the previous year's stubble remains in the ground, are still being argued, but the method is being used increasingly.

The huge plows that the Lewtons use for preparing ground cut into the soil only about 6 to 8 inches. Because of the repeated use of the plows, the soil just below where the bottom of the plow rides becomes hard enough to inhibit moisture flow into the subsoil and forms what is called "hardpan." To break this up, the Lewtons have bought subsoilers, monster plows that go down about 22 inches, breaking up the soil that has formed the

Young winter wheat needs to germinate rapidly in the fall so it can get above the ground before freezing temperatures stop its growth. It protects the soil from blowing and washing. Much wheat ground, because it is in areas of minimal rainfall and little ground cover, is particularly subject to wind and water erosion. Government programs are endeavoring to provide incentives to get much of the dry, erodable land out of production.

hardpan. They subsoil each field about every four or five years.

The Lewtons are hoping to experiment with sunflowers as an extra crop, both to diversify and to help break up the hardpan. "Sunflowers have roots that go down maybe 3 or 4 feet, and while they're going down they break up the hardpan," Curtis explains. "So they're good for that. But the price for sunflowers right now doesn't justify planting them. I figure the price is so bad that maybe a lot of people won't plant them this year, and the price will go up. So maybe we'll take a chance on a few hundred acres."

Wheat usually is planted with a piece of equipment called a drill. Other grains, such as corn, soybeans, and sorghum, are put into the ground with a planter. A drill may plant as much as a 72-foot swath in a single pass, although most drills are smaller. As it is pulled behind a tractor, the drill opens a series of continuous thin slits into the ground, drops seeds in at regular intervals, then closes the slits. Seeds may be dropped 1 to 6 inches deep into the soil.

Generally, the ground must be fertilized first, although some years the Lewtons find there are enough nutrients left from earlier years to make fertilizing unnecessary. Much of the fertilizer for wheat is applied as a white gas from a mobile tank containing anhydrous ammonia, a form of nitrogen. Before planting, a tractor pulling a cylindrical tank passes through the field. Knives attached to a bar behind the tractor cut into the soil and the ammonia gas is injected into the ground and clings to soil particles underground. The nitrogen is released gradually. Fertilizer can also be put on as a liquid, which is sprayed or dribbled onto the ground, or as a solid, in powder or granules.

In eastern Colorado, where Curtis Lewton and his family farm hard red winter wheat, planting starts in late August. "Generally we start on the corporation ground, the land the family farms jointly. The soil is a little bit lighter than the rest, and we want to stop it from blowing by getting the wheat above ground as quickly as we can," he explains. "The private ground each of us farms is more solid, and so we go to it next. We aim to plant early enough so that all the wheat's up and protecting the ground before freezing stops the growth for the winter."

The half of the ground that is left fallow conserves moisture for the crop the following year. Although not planted, the fallow ground must be cultivated, because if left untilled, it will sprout weeds, and weeds use up precious moisture. There are various tools for this cultivation. The Lewtons use moldboard plows for part of it, plowing just deep enough to tear up the weeds, maybe 5 to 7 inches.

Curtis Lewton defends the plowing: "The college says that's not the way to farm. They're encouraging chemical fallow these days—that's very big now—but we find here on our ground that if we use chemical fallow, we eventually get hit with cheat grass. Now cheat grass, it's hard to control unless you moldboard and turn that seed under, deep enough so it won't come through."

He keeps a close eye on weeds. "We spray pretty much every year for weeds," he explains. "Funny how weeds move in some places. That ground we farm out east, there's quite a bit of pasture around it, and antelope carry spikeweed in on it so bad it's hard to

Much of our major wheat country is so dry that a crop is produced only every other year. During alternate years, the fields are "fallowed"—left unplanted—but tilled to keep down weeds. This conserves moisture for the crop year but increases the probability of erosion on the unplanted ground.

keep up with it. We spray, but we've also been experimenting with minimum tillage out there, to see if it helps."

As wheat gets taller during the summer, moving equipment over the fields becomes less possible and aerial application becomes the satisfactory alternative. Airplanes "fly on" not only herbicides to control weeds but insecticides to kill off obnoxious insects, and often they fly on fertilizer as well.

Insects pose a threat to wheat that varies from year to year and from area to area. Grasshoppers can sweep through a field and leave most of the grain on the ground, chewed into uselessness. It's possible to see as many as a dozen grasshoppers on a single stalk.

Harvesting

When the wheat has finally ripened—received enough water and fertilizer, survived the insects and the weeds—it's harvest time. To the public, harvest is the most romantic time for wheat. To the farmer, harvest is damned hard work. The entire process must be performed rapidly, when the wheat is dry enough but before inclement weather damages it. The combine is the magic instrument for this arduous operation.

A combine is a wondrous machine that wheezes, purrs, roars, thumps, bumps, and glides. It is an astonishing collection of gears, belts, pulleys, fans, and cylinders—all of

Each of these wheat combines cuts a swath **24** feet wide, and together they can cut almost a square mile a day. A combine cuts the wheat, separates the grain from the chaff, pumps the grain into a storage tank on the combine, and throws the chaff back onto the field.

Combines unload the grain into trucks waiting to haul it to storage. Frequently, to speed up the process, the trucks roll alongside so the combine can continue cutting while unloading.

which seem to move at odds with one another but actually mesh together superbly.

The jobs the combine performs are simple enough: it cuts the standing stalks of grain, moves them into its interior works, separates the kernels from the chaff and stems, deposits the kernels in a storage bin, and dumps the unwanted parts back onto the ground, either in a smooth line following the combine or broadcast across the field. It does all this while moving rapidly across a wheat field. If it were human, the combine would do justice to a Charlie Chaplin movie!

A John Deere 8820 combine weighs 11 tons without any auxiliary equipment; it can be equipped to cut a 30-foot swath of grain and will combine 150 acres on a good day. At 40 bushels to the acre, that's 6,000 bushels of wheat in one day! Having paid close to $100,000 for one of these monsters, the owner naturally feels entitled to ride in an air-conditioned cab, seated on a carefully padded seat, his stereo turned on high, driving with a tilt-wheel steering wheel, while he rolls across the field. Harvest techniques have come a long way, and their progress is typical of the ways in which capital—combined with ingenuity—has replaced labor in farming.

At some time in past history, presumably the earliest agriculturists simply cut the grain with a knife (which was probably chipped from stone), stomped on the kernels while they were still attached to the stalks, and then separated the kernels by hand. Sometime very early they surely also discovered that after breaking the kernels and the chaff apart, they could toss them in the air and, if the wind was right, the chaff would blow away while the heavier grain dropped to the ground or into a container. Called winnowing, this method is still used in many parts of the world to separate the wheat from the chaff.

Harvest techniques improved slowly. The curved sickle was a great improvement over the straight knife and was used for 4,000 years. The scythe, with its long, curved wooden handle, was faster than a sickle and a lot easier on the back. A cradle—a light wooden framework attached to a scythe—collected the stalks as they were cut, so they could be dumped into a pile, making it easier to pick them up as a group. But so far all the work was done by human labor.

By the 1830's, an energetic, imaginative man named Cyrus McCormick eliminated the hard labor required of a scythe by inventing the horse-drawn reaper. With later improvements, the reaper was capable of cutting the grain, putting it into bundles, tying the bundles, then dropping them onto the field. These sheaves were gathered together and stacked to dry before they were hauled away to be threshed. The reapers and reaper binders were manufactured in larger and larger sizes that required more and more horses to pull them. It became a common sight to see fifteen or twenty horses pulling a single piece of equipment. Throughout the era of the reaper, however, the grain still had to be taken from the field to be threshed.

Threshing—the separating of the grain kernels from the rest of the plant—progressed from peasants stomping on the grain to break the kernels loose, to horses and cattle stomping on the grain (the farmer got free manure with his wheat). Also in common use was the flail, which consisted of two yard-long pieces

Kansas is the major wheat producer, with North Dakota, Texas, and Oklahoma a distance behind. Some wheat is grown in almost all states, but the crop produces a smaller return per acre than most other crops. The greatest production, therefore, is limited to areas where, because of a lack of moisture, other crops are at a disadvantage. Wheat can be grown very well under irrigation, but more valuable crops can also, so most wheat is planted in dry country.

of wood linked together with a rope or leather thong. Gripping the end of one club, the operator would swing it so that the other club flailed to beat the grain free. Primitive, but no manure.

As coal-fired boilers brought steam power to the farm in the early 1900s, threshing became mechanized, and giant machines replaced hand-threshing techniques. Even these advancements failed to relieve the inefficiencies of the harvesting process: cutting and leaving the grain stacked in the field, later carrying it off the field into a threshing area, then moving the chaff back into the field for disposal. This way of harvesting remained unchanged until about sixty years ago, when the combine came into being.

The combine does just what its name suggests: it "combines" the jobs of cutting the grain and threshing it, and it does both jobs while rolling across the harvest field. The kernels are trucked off the field, but the unwanted chaff is simply dropped back onto the ground.

A combine may cost a pile of money, but look at what it replaces. One man using a sickle could cut less than an acre a day, leaving the grain on the ground. Another worker had to gather the grain and tie it. Still to be done was the labor of threshing. One acre was a lot of cutting with a sickle or scythe, romantic as it may look in old paintings. An experienced farmhand operating a flail could break out less than a bushel of grain in an hour, and someone else had to do the job of winnowing. No wonder early grain harvesting required large families, endless labor, and lots of luck to get the grain out of the field before it rotted.

Wheat harvesting begins in late May in Texas and continues until September in the states along the Canadian border. The entire process centers on whether the wheat is "ready." Moisture testing is necessary. If the wheat's not acceptably dry, the storage elevator won't accept it because it won't keep. Usually grain that contains more than 14 percent moisture can't be safely stored, and so the combine crew waits until the moisture drops below that level.

Most big grain farmers have two-way radios in their rolling equipment, and as soon as the radio speakers blare "Test's OK, let's go!" everyone breathes a sigh of relief and moves into high gear. It's rare to start early in the morning, because dew pushes the grain's moisture level too high to allow cutting, but by nine or ten o'clock the combines are edging into the fields to make enough of a run for a test sample. From the first "OK!" it's then a long day, or a long day and night, for the combines will likely keep running until the next morning's dew forms and builds up moisture, shutting them down. That may occur as late as midnight or two in the morning.

Meals are eaten along the edge of a field, out of the back of a pickup, or in the combine cab while the combine is cutting—a bumpy sort of dining table but part of the freneticism of getting the wheat in. This hurry doesn't come from some ridiculous tradition; it comes from knowing that your entire year's labor and capital input are sitting out there, vulnerable.

A few years back, I watched as a pair of grain combines rolled to the edge of a wheat field in eastern Colorado, ready to start cut-

In warmer parts of the country, double cropping is often possible. Some quick-maturing varieties of small grains, particularly barley, can be planted in the fall and will get a good enough start before the winter to make a crop early in the summer. A second crop, often soybeans, can be planted in time to be harvested before the winter freeze.

ting a test strip of ripe wheat. The sky was dark, threatening, and the farm owner kept looking upward while he talked with the combine crew. Suddenly the weather became more than threatening. As a hailstorm swept over us, we jumped back into a pickup truck for protection from the stinging pellets. Within minutes the storm had passed and we walked the field, which was covered with marble-sized hailstones. In only ten minutes

the grain had been knocked from the stalk and was lying on the ground, irretrievable, a complete loss. The farmer shrugged his shoulders and directed the combines to another field a couple of miles away; maybe luck would be better there.

For a few years in the 1970s, the price for wheat was high enough that if a farmer made it one year he could get by with a disaster the next. That's no longer the case; one bad year and the bankers are at the door. There's plenty of reason to get the crop off the field and into storage.

One of the romantic figures of the grain harvest is the custom combiner. This modern-day gypsy really isn't that at all. He's a hard-working agribusinessman whose workplace happens to move from one part of the country to another. Floyd Oliver, who lives near Amarillo, Texas, is a custom combiner. He started working on a combine crew when he was twelve and has run his own combining operation for thirty-five years. He has been cutting wheat on some of the same farms for the entire time. His combining year starts in late May when he and his wife and crew take to the road, starting with fields 150 miles east of them in Texas, then moving north through Kansas, eastern Colorado, and Wyoming. By the end of August they are in Plentywood, Montana, just a few miles from the Canadian border, having combined grain all along the way.

His equipment includes three John Deere 7720 combines with 24-foot headers, four trucks, a bus outfitted with seven bunks and a sunken tub for the crew, a 40-foot trailer with kitchen and living quarters for the cook, and a motor home for Floyd and his wife. His overall investment is more than half a million dollars.

Floyd estimates he combines 10,000 acres of wheat a season, and when he finishes with wheat, he harvests an additional 2,000 acres of sorghum as it ripens in October near his home in Texas. Each of his combines cuts about 15 acres an hour, so he can harvest about 450 acres a day—although it depends on the wheat. Heavy production slows the combines; poor wheat can be cut faster.

Storing and Shipping

The Lewtons combine their own grain and store most of their wheat at home, in their own bins; it moves off their farms only when the wheat is sold. But many farmers use commercial storage. Commercial grain elevators dot the horizon of the grain states, varying from two or three silos for the "country" elevators to giant spreads for the "terminals." They are often called prairie castles.

The biggest concentration of grain elevators is in Hutchinson, Kansas. From the air the eastern section of the town looks like a collection of stretched-out skyscrapers. One of the largest elevators in the world is there—Union Equity's Elevator B—which is almost half a mile in length and holds 18 million bushels of grain. The controls in these elevators are so precise that wheat can be stored both by quality and by variety. Low-gluten wheat (gluten's the stuff that makes bread dough elastic so the final product will have a fine texture) can be kept apart from high gluten. In fact, the building is so huge that, in the gallery running across the top of the silos, operators ride bicycles when they check the bins and belts.

Most of the wheat is destined for export; in 1984 about 60 percent went out of the country. More wheat leaves the United States from New Orleans than from any other port, and much of that supply arrives by barge down the Mississippi. Barge transport is slow but cheap. The port of Minneapolis-St. Paul, at the head of the Mississippi, loads more than 5 million tons of grain each year for shipment downriver, mostly to New Orleans.

Barges haul immense loads. A semi-trailer on the highway can load about 800 bushels

Much of our western wheat, grown mostly in the Palouse area of Washington, shown here, is sent from west coast ports to Japan and South Korea, both major importers. Usually more than half of our wheat is sold outside the United States, with New Orleans being the largest grain port.

of grain, and a fifty-car freight train using "big John" cars can haul about 100,000 bushels. But 100,000 bushels will go onto just two Mississippi barges, and one towboat can handle twenty barges, a total of 1 million bushels!

Grain from the Midwest also goes out through the Great Lakes. Duluth-Superior is a major port, and it handles both the "salties," which take grain overseas through the St. Lawrence Seaway, and the "lakers," which move the grain from one Great Lakes-port to another. Recently I made arrangements to visit a ship while it was loading in Superior. I talked with the captain while the wheat poured into the ship's hold from huge

tubes attached to storage elevators on shore.

The captain explained to me that his German-owned ship was registered in Singapore, had a first mate who was Greek, and a crew that was mostly Spanish. "We were commissioned in Germany; I came aboard in Belgium; we went to South America to pick up a load of soybeans for Gdynia, Poland, then came here in ballast. We're taking on 15,000 tons of durum wheat here; we'll go out through the lakes and the Seaway and deliver the wheat to Algeria; and after that, well, we'll have to see."

It's a long way from Curtis Lewton's wheat fields in Colorado to a market in Algeria.

Crop Storage

Crop storage has become too sophisticated and too specialized to be confined to barns, and as we have developed great surpluses of grain, increasing numbers of storage facilities have moved off the farm entirely. Metal and concrete have largely replaced wood as materials for storage containers, and plastic is becoming a very common farm storage material.

Most of the big old barns were big because they were made for storage of loose hay. Most hay today is stored under compression in bales or compressed stacks, allowing it to be kept outside with minimal protection, and to occupy far less space.

Thanks to our increased ability to produce ever-growing surpluses, and to move grain cheaply, grain storage facilities have become more concentrated, and larger. There's a joke among agricultural economists that says the surplus problem will eventually be solved when the amount of land needed for storage facilities will cover enough of the land previously available for grain protection to do away with surpluses—from then on the surpluses will be managed by constructing, or tearing down, storage facilities.

The storage of blockbuster crops is one of our major agribusiness activities. Almost half a mile long, Union Equity's Elevator B in Hutchinson, Kansas, (top left) holds 18 million tons of grain. Most commercial grain elevators are now concrete. Potatoes are often stored in steel quonset huts usually still referred to as "potato cellars."

New grain storage techniques involve loading the grain into conical piles, usually with ventilating pipes spread through the base, then covering it with heavy plastic—cheap but effective. On-farm storage of grain is still mostly in steel bin. In dry country, plastic covers, such as this one on cotton, often provide adequaate protection.

Corn

The trouble with corn is that there is too much of it. Otherwise it's a great crop. It grows fabulously well; it's suited to much of the country; it responds well to irrigation techniques; technology has been able to increase yields almost unbelievably; and farmers love to grow it. Uncle Sam even seems to love to pay for its being grown. The trouble is mostly how to get rid of our immense production.

We've been growing corn on about 60 million to 70 million acres, and our total annual production is often in the neighborhood of 7 billion to 8 billion bushels, with a value likely to run about $20 billion. Although there's some corn grown in every state in the continental United States, Iowa and Illinois account for more than a third of the total crop, with Nebraska tagging along behind. Some-

thing more than 10 percent of the crop is grown on irrigated ground.

Lest you wonder how we consume so much corn on the cob, don't. The corn we eat is "sweet" corn, and it occupies only about 1 percent as much land as the crop that is usually referred to as commercial corn or field corn or just corn, which is used primarily for feeding animals and for industrial purposes—high-fructose corn syrup, ethanol, and dozens of other manufactured products. In addition to sweet corn and commercial corn there is also seed corn, a highly complicated crop that generates the seed for both sweet corn and commercial corn.

Production of corn has grown astonishingly during the past fifty years. In 1935, prior to the use of hybrid corn seed, the average production per acre was 24 bushels; in

Corn is the most valuable crop, with the annual value likely to be about $20 billion. Like the other blockbuster crops, it is grown on about 60 million acres—although the acreage is decreasing because of the government's farm programs designed to cut down on surpluses.

1970 it was 72 bushels. Recently average production has consistently been topping 100 bushels, and it can do even better. Many farmers have fields that yield 200 bushels, and some have even achieved 300 bushels to the acre. Under current conditions, however, while aiming for higher production per acre may be possible, it may not be economically sensible and could be disastrous to an already stressed marketing system.

Sit down to talk with Dennis and Linda Carlson, and you quickly understand they are a couple whose current way of life is closely tied to the success of corn. Denny farms in partnership with his father, two brothers, and a brother-in-law. His grandfather, whom he calls Gramp, came to the United States from Sweden to escape World War I, only to be drafted into the American army. After the war he began to farm corn in Iowa. Linda grew up on a farm a few miles from where she and Dennis now live.

The Carlsons raise 500 acres of commercial corn in rotation with soybeans on excellent

Iowa land south of Fort Dodge. All the commercial corn they produce, as well as additional corn they have to buy on the open market, gets fed to two hog operations the family partnership owns. They also grow and sell seed corn for the Garst Seed Company of Coon Rapids, Iowa. That's being involved with corn.

The commercial corn that Dennis Carlson grows is hybrid corn, and the seed he raises provides the seed for the commercial corn. The hybridization of corn is often called one of the modern miracles of agriculture, which it assuredly is, and it is largely responsible for the great increases in the production of corn over the past fifty years. Essentially, hybridizing utilizes "inbred" lines, corn varieties that for many generations have been bred with themselves. Inbreeding may result in creating very desirable characteristics but it also results in poor production. By crossing these inbred lines, "hybrid vigor" results.

In "double crosses," four inbred lines are crossed. Let's say line A is crossed with line B, and lines C and D are crossed with each other. Then the two crosses produced by these matings are crossed, getting ABCD. While the double cross was used in the early days of hybridization, today breeders have gone mostly to a single cross, utilizing only two inbred lines, since over the years inbred lines have become much more even and dependable. The result is corn that is highly productive, very uniform, the same height, with ears the same distance off the ground, simultaneous maturity—all characteristics very similar, and all weakness or resistance to disease and growing conditions also the same.

Each cross in a seed company's line is numbered for identification, and its characteristics are thoroughly outlined. Farmers search for a cross that is especially suited to their specific conditions of climate, soil, insects, and other variables. Their advisers are likely to be the seed corn salesmen, who are usually corn growers themselves.

Growing Seed Corn

In addition to their 500 acres of commercial corn, the Carlsons raise almost 400 acres of seed corn, which is a different sort of crop from commercial corn. The seed company contracts for it, with payment on a complicated schedule that takes into account the fact that seed corn produces less than half as much seed as commercial corn.

Seed corn also takes up a lot more ground than commercial corn. "It has to be raised in isolation from other corn," Denny explains; "otherwise it might cross-pollinate with ordinary corn. We raise it on the home farm down at Gowrie, southwest of here, where we can keep a wide belt of unplanted ground around it."

The Carlsons have a twelve-row planter, which they use for both the seed corn and their commercial crop. They plant two varieties of seed corn, one of which will eventually be used as the female, and the other the male. They usually plant two male rows for every six female rows.

Corn has both male and female parts on each plant, making it possible for the corn to pollinate itself. It also can be pollinated by pollen blowing from nearby fields. That's why the seed corn takes a lot of ground. The male part, the tassel, develops at the top of

About four times as much corn per acre is grown today as was grown fifty years ago, partly because of hybridization, but also because of the increased application of fertilizers and more sophisticated growing techniques.

the plant and produces the pollen. It's estimated that a typical plant produces between 2 million and 5 million pollen grains. The female receptor for the pollen is the silk, which on a mature plant is the brown hairlike stuff at the end of the ear.

Grains of pollen, released from the tassel into the air, drop on the silk and penetrate each strand, then move down the interior tube of the silk to what becomes the ear, and a seed, or kernel, grows from the silk strand. An ear of corn has 600 to 1,000 seeds.

When corn is being raised for breeding, the variety that is to be the female corn has to be prevented from producing pollen of its own— the pollen has to come from the other variety. The tassels on the female corn must be removed after the tassel has developed but before it begins shedding pollen. Because seed corn breeding stock varies greatly in its growing habits and the tassels ripen at different times, the detasseling has to be done in several stages.

This is the responsibility of the seed company, and Garst hires more than 15,000 helpers for a short time each summer to detassel

Corn planting is completely mechanized, often with eight or twelve rows planted at one time. Chemicals, including fertilizers, insecticides, and herbicides, are frequently loaded into containers attached to the planter and applied automatically as the corn goes into the ground.

Anhydrous ammonia, a common source of nitrogen, is a gas that is stored under pressure in wheeled tanks. As the tanks are pulled behind tractors through the field, the gas is knifed into the ground under pressure and clings to soil particles, from which it is taken up by plant roots.

its 60,000 acres of seed corn, including Denny Carlson's. The corn also has to be "rogued" by technicians who frequently walk the fields checking it. Breeding corn sometimes has throwbacks, or rogues, plants that revert to earlier generations and different types of growth. These plants are removed from the field by hand so they don't rebreed into the product that goes to the final grower.

Back in the 1960s an alternative to the destasseling chore was discovered. A race of corn in which the male part was sterile was developed and enthusiastically accepted by the corn breeding companies. The one widely used, on about 80 percent of the corn crop, was the "Texas" type, or Texas T as it was called. It offered an immense savings in labor costs because it didn't require manual detasseling. All went well until 1970.

That year—for reasons that still aren't clearly understood but must have certainly involved weather conditions—a new race of the southern corn blight, called race T, developed. Endemic in the South, southern corn blight had always been a nuisance but not a major problem. The new race was something else. It could withstand cooler weather, and it moved from the South into the corn belt. Corn that had been bred using the Texas T cytoplasmic male sterility was highly susceptible to the new T type of southern corn blight. While agronomists still argue over how much corn was lost, it seems likely this loss amounted to 10 to 20 percent of the total crop—a near disaster.

Remarkably, by the following growing season, seed corn that was resistant to the T race of southern corn blight had been produced for the growers, but the newly resistant crosses required detasseling. Development of the new crosses was sped up by growing an extra generation of seed in South America,

Of an 8 billion bushel corn crop, close to 2 billion bushels may be sold abroad, an amount that varies greatly from year to year. Unlike wheat, which is used throughout the world, corn is a crop of limited interest to other countries. Japan has been the biggest importer. China once was a major importer, but its own corn production has soared so that now it is second to the United States as a producer. The Eastern bloc countries vary their buying greatly from year to year, depending on their own crops and on the political climate. The Soviet Union has usually been second to Japan as an importer, but its imports vary.

Export tonnages seem to be diminishing, and the amounts fed to animals are moving up only gradually—up only 15 percent in the past fifteen years. The amount of corn used in food and industrial products, however, has soared in fifteen years from 400 million bushels to 1.2 billion bushels, and the use curve may be getting even steeper. Corn is a product that, like petroleum, can be put to almost unlimited uses, and those uses are changing rapidly and dramatically.

The biggest jump has been in the production of high-fructose corn syrup (HFCS). Almost unknown in 1971, when it used only 10 million bushels of corn, in 1986 it required 320 million bushels to meet the market demand. Where liquid sugar is used, HFCS has just about taken over the sugar market. It is derailing sugar imports and cutting into domestic production of both beet and cane sugar, another example of one agricultural crop, corn, competing with another agricultural crop, sugar. Not surprisingly, the beet and cane sugar producers want government

If corn is harvested for grain when the moisture is high enough to make spoilage in storage a possibility, it can be dried to a predetermined moisture content by the application of heat and air.

protection not only from imported sugar, but from HFCS.

The other fast-growing product made from corn competes not with another crop but with petroleum. Ethanol—alcohol made from corn—began appearing in the headlines during the oil crisis of the 1970s, when it was touted as a salvation because it could be used to extend gasoline. "Gasohol" became the magic that would save us from the political crisis, and retail taxes on ethanol in gasoline were dropped as an incentive to use less petroleum and more ethanol—which is considerably more expensive than gasoline. Enthusiasts, mostly corn growers looking for an outlet for their overproduction, built ethanol refineries around the country. The petroleum crisis ended, and so did much of the demand for ethanol. Many of the small refining operations went broke and closed down, in spite of the tax advantages for ethanol production.

The positive effects of ethanol in gasoline have recently returned to the headlines. Because it is believed to burn more cleanly than gasoline, a number of cities began to regard it as one way to alleviate their air pollution problems. The phasing out of lead in gasoline also created interest in ethanol. Other additives can be used to minimize pollution from gasoline, possibly more cheaply than ethanol, but it is certainly in the running. Lurking in the corn refiner's bag of tricks is, actually, the solution to the farm problem. This is my own theory, of course, but it hasn't been tried and almost everything else has. It may not be practical, but it's simple. We need a one- or two-sentence law, if there is such a thing, that says: "The use of ethanol in gasoline will be phased in over a five-year period, in such a way that at the end of five years all gasoline, or gasohol, used in automobiles will contain 10 percent ethanol."

The figures look roughly like this. We use about 120 billion gallons of gasoline annually in the United States. One bushel of corn makes 2½ gallons of ethanol. If we replaced gasoline with gasohol, which is 90 percent gasoline and 10 percent ethanol, we would use 12 billion gallons of ethanol. That would consume almost 5 billion bushels of corn, which is more than half of our annual production. Soon we would have a corn shortage, which would draw acres from other surplus crops into corn production and eventually balance out so there would be no surplus of anything. We could, indeed, let our farmers plant as much as they wished.

The gasohol would cost more than gasoline, but taxes could be lowered because we'd have no farm subsidies to pay—about $20 billion to $30 billion a year. If we passed another simple law putting an import tax on foreign oil, an idea that has been widely suggested for years, we would not kill our own oil industry but would greatly help our balance-of-trade problem.

Ah, but that's too rational, too simple a solution. Or is it?

Soybeans

There's hardly any crop less glamorous than soybeans. Wheat waves majestically in the wind, corn is tall and stately, cotton bolls opening seem infinitely productive, peanuts taste good, but soybeans—they just grow. Producing hairy, dark pods almost hidden by foliage, the skinny plants grow maybe 3 feet high, often falling over one another in disorderly fashion. It is, indeed, an undistinguished crop.

Soybeans do have some things going for them, though. Because they are legumes, they generate nitrogen fertilizer not only for themselves but for the next crop following them. There are other legumes widely used in farming, alfalfa being the one that comes most frequently to mind.

Soybeans also have the advantage of being, like corn, convertible into other products, everything from plastics to cattle feed. A bushel of beans, which weighs 60 pounds, converts into 10½ pounds of soybean oil and 45 pounds of meal. The oil goes into margarine, salad oil, shortening, and dozens of other food products. It also has industrial uses in such products as paints and varnishes. Like corn, once it has been converted into some basic chemicals, the uses to which it can be put are more related to economics than to chemistry.

Until a few years ago, soybeans had another remarkable quality: they were in short supply in the world markets, which spurred production in the United States. We produced our first 1 billion bushel crop in 1968 and by 1979 had topped 2 billion bushels. A combination of greater acreage and higher prices moved the soybean crop's value from a little over $3 billion in 1970 to almost $14 billion in 1980; the value has been dropping

Soybeans are a relatively new crop for the United States. They did not come into heavy production until the end of World War II, and then for many years were in the happy position of having a greater demand than supply. Those days are gone as international production, particularly from Brazil, has soared. Our own acreage is more than adequate for domestic use, as well as export sales.

Soybeans can be grown in various ways. They can be planted in wide rows 30 inches apart, but they can also be planted much more densely with a grain drill or even from the air. In rows they can be cultivated to keep down weeds, but in random denser plantings the soybeans must be thick enough to smother weeds.

When small, soybean plants must be protected from competition with weeds, either by herbicides or by cultivating. When they mature, soybean plants have a tendency to blow down and become difficult for the combine head to pick up.

off since then, both in acreage and in price. The reason for the drop is that the rest of the world, particularly Brazil, has been catching up with the good thing the United States had. Brazilian technology, helped by equipment and know-how partially imported from the United States, has made that country's production soar, and its presence in the world market has put prices under pressure. The third-largest producer is China, which is also likely to add pressure on world prices as its technology improves and it produces more than it can use.

Soybeans are a relatively new crop in the United States, but their history in China dates back at least 3,000 years. In China they are still consumed primarily as human food; their high percentage of protein makes them a nutritional bonanza, and their protein is far cheaper than meat. In the United States practically all soybeans end up with their oil extracted and the remaining meal going into animal food. Soybean meal is more than 30 percent protein.

Although soybean plants were brought to the United States early in the nineteenth century, it wasn't until after the First World War that there was any interest in them as a crop. An agronomist named W. B. Morse brought some 5,000 samples of seeds back from China to test here, but there was little interest in the crop. Then in World War II a shortage of edible oil caused production to double. Following the war, our animal production increased greatly, and along with it a need for feed. Also the American Soybean Association began to promote the use of soybeans abroad, which it is still actively doing today.

American agriculture is full of organizations interested in promoting foreign use of our crops. Most of them are at least partially funded by a "checkoff," a percentage of all the sales of a product, which goes to the organization for research on improving the crop and for promotion of the crop. The American Soybean Association's activities in Turkey provide a good example of what an organization can do. In the early 1980s the association worked with the Turkish government to lower that country's import duty on soybeans and make soybean products available to Turkish livestock and poultry producers at a feasible price. Then it worked with the grain-handling and feed-manufacturing industries in Turkey on modernizing facilities. Finally, it began a series of seminars for poultry producers to show them how to use soybean meal in poultry rations.

The association sponsored poultry-feeding trials, publications, and even video presentations. It helped bring twenty-eight Turkish feed-manufacturing technicians to a week-long course at Kansas State University. Poultry production in Turkey is projected to double within the next five years, and the American Soybean Association is hoping that the American soybean industry can sell them the soybeans for this expansion. The estimate is that Turkey might use as much as 36 million bushels of beans within a few years.

Cultivation

One of the reasons for the soybean's success is that it can be grown under a multitude of conditions. Illinois is the major grower, followed closely by Iowa, but Minnesota to the north and Arkansas to the south are also important producers. The heavily irrigated

Soybeans are usually planted on finely tilled ground (left), but often as a second crop they are planted directly into the stubble of an earlier crop, such as barley. Because they leave a lot of unshaded open ground, soybeans are bothered by competition from weeds. Modern herbicides differentiate between soybeans and weeds and can be "broadcast" on the ground to keep weeds down. The least costly way to use herbicides is by "zapping" weeds after they appear, using rigs (below) driven through the fields. The herbicide is thus used only where weeds have appeared—at less cost and with fewer environmental consequences.

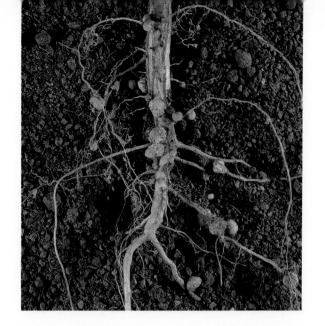

Soybeans are legumes. Legumes can separate nitrogen from the air and absorb it through nodules attached to the roots. Some nitrogen remains in the ground after soybeans are combined, for use by the next crop. Tests show that corn that follows soybeans yields about 10 to 15 percent more than corn that follows a nonleguminous crop. Scientists are endeavoring to transfer the nitrogen-fixing ability of legumes to other crops such as corn that are not naturally leguminous.

states produce little—their water goes for more valuable crops.

Beans can be planted in standard rows, like corn, or they can be planted much more densely. When corn is planted at much more than 22,000 seeds to the acre, production decreases sharply. Soybeans, on the other hand, are routinely planted at five or six times that density. Being shorter and having less foliage, they shade themselves less. Whereas most corn is planted in rows that are 30 to 40 inches apart, soybeans are often planted in rows 10 inches apart. Soybeans are usually planted with the standard row planter also used for corn, but they also can be "drilled," as wheat is, or even seeded by airplane, as rice is.

Before soybeans are planted, they are usually treated with an inoculant. The bacteria of the inoculant stimulate the growth of nodules on the roots of the plants, which "fix" nitrogen into the soil, a valuable characteristic of legumes. Soybeans are thus an excellent crop to rotate with corn because they

supply part of the nitrogen for the corn crop that follows them. Tests show that corn following soybeans yields about 10 to 15 percent more than corn following corn.

While it is very common nowadays for corn to follow corn, the practice allows insect and disease problems to build up in the soil. Introducing soybeans into a rotation system gives the soil a year's rest from the corn problems and introduces additional nitrogen. Rotating is still the standard cropping technique through most of the Midwest, although there are many farms that plant corn year after year and cope with insects and disease through the sophisticated use of chemicals.

Soybeans are also grown frequently as a second crop. An early grain, such as barley or rye, can be grown and harvested—in the Midwest by mid-June, in the South even earlier—then short-season soybeans planted, usually directly in the stubble of the grain crop. Planting in stubble minimizes one of the weaknesses of soybeans. Because they are short plants, have a small root structure, and grow relatively little foliage, and because their roots loosen the soil heavily, they can cause erosion. Fields planted to beans, particularly when they are still small, are subject to heavy erosion damage from rain. Therefore, if the grain stubble can remain in the ground while the beans are getting established, the possibility of erosion is minimized.

Weeds are a bigger problem in soybeans than in corn. It takes beans longer to "canopy" over and shade the ground; once the ground is shaded, weeds have little chance to grow. Herbicides are therefore used extensively with soybean production. They can be put onto the soil before the beans are planted, sprayed in the planted rows behind the planter, or applied after the beans are up.

When soybeans follow corn, there is a tendency for "volunteer" corn to appear, sprouting from seeds knocked into the ground from the previous year's crop. When the corn gets taller than the beans, an herbicide can be applied to it by a "rope" or "wick" applicator mounted horizontally on the front of a trac-

tor and set high enough to clear the tops of the beans. As the tractor moves through the field, the wick is dragged against the volunteer corn but doesn't touch the beans. Very little chemical is needed.

Another technique uses a tractor that usually has four seats mounted on it to carry four operators equipped with spray nozzles. As the tractor moves through the field, the spray operators look for any weeds and zap them. Again, very little herbicide is used. Ingenuity has been put to work reducing costs while also minimizing the amounts of chemicals placed in the field.

Illinois and Iowa are the two leading producers of soybeans as well as corn. In 1985 their soybeans brought in $3.6 billion; corn brought them $7.7 billion. Soybeans may be the less glamorous member of the team, but together corn and soybeans represent a remarkable part of agricultural income in the midwestern agricultural belt. They give rise to the affluence that has led to what has been called CS&F farmers: corn, soybeans, and Florida.

Forage and Hay

In their unpretentious way, forage and hay are blockbuster in size. Without them there could scarcely be an animal industry since they provide the basic food for cattle and sheep. More than 60 percent of the world's farmland is in meadows and pasture. In the United States 60 million acres are planted to hay, and at least several hundred million acres are used for pasture—so much that it's difficult even to calculate.

What, exactly, *are* forage and hay? Forage is simply any crop that animals graze. They graze forage on pasture. In the past forage has mostly been native grasses, but today it has become more sophisticated. Forage crops have changed from grasses alone to include a great variety of other plants, mostly legumes such as alfalfa and the clovers, but also young grains such as wheat and barley.

Hay is dried forage, and alfalfa has increas-ingly become the most important of the hay crops. It accounts for about half of the hay produced, although in the drier western areas and the Rockies native grasses are still important. Whereas animals usually go to forage for grazing (although that is changing somewhat), hay is brought to the animals.

Straw is not to be confused with hay. Straw is the stem of grains. After the grain has been removed, the straw is either discarded and plowed into the ground or baled and used for animal bedding. It can be eaten by animals, but only incidentally, since it has limited nutritional value.

Silage feeds animals also. Silage is plant material that is finely chopped, compressed, and packed tightly in storage silos to exclude air and permit slight fermentation. Corn is the major silage crop, using about 10 million acres of land, but silage can also be made from standard forage crops. In the past thirty years a combination of hay and silage, called haylage, has been developed and is widely used. Made from forage crops, haylage is chopped in the field and packed like silage into silos or other air-excluding storage.

The animals that can make the most of forage, hay, and silage are ruminants like cattle and sheep. Ruminants have a complex series of stomachs that allow them to convert forage crops into protein. Bacteria in their first stomachs aid in digesting the forage crops, building protein in the bacterial cells. Stomachs farther along in the digestive tract then digest the bacteria. That's what allows roughage like grass to be converted into valuable feed.

Pasture

Most of us tend to think of grasses as plants that just grow naturally, untended, and once it was almost that way. Today, however, forage crops are much more varied, and their production is often as carefully managed as any other crop. To understand the change, we need to take a historical look at how forage developed.

From the Dakotas south through Oklahoma, before the prairies were settled, the untilled soil annually produced remarkably heavy crops of grass. Reports by early explorers usually included some reference to "buffalo hidden by the tall grass" or "buffalo up to their bellies in the lush summer pasture." Many of these early references were to the Flinthills area of central Kansas, still a great cattle-grazing area. It is still a considerable mystery to soil scientists how the virgin grass managed to grow so well.

The tallgrass prairie ecosystem, simple and beautiful as it looks, is immensely complicated, the result of interactions among the soil, climate, topography, fire, and creatures that live on the prairie. When these elements change, the system changes. If there is no fire, for example, forest gradually invades. In earlier times, frequent fires were caused by lightning and by Indians. The trees were killed off, but the prairie plants recovered rapidly from the burning. Bison, elk, and pronghorn grazed the prairies, although it's difficult to know in what numbers and how intensively because they moved easily from one area to another as range conditions changed.

Virgin prairie has almost completely disappeared, victim of both agricultural and urban development during the last century, but a few scattered pockets of original prairie grass can still be found. Probably the most interesting is the Konza Prairie Research Natural Area, just south of Manhattan, Kansas, where more than 13 square miles of remaining tallgrass prairie, mostly bluestem grass, are being studied.

The research at Konza today is being carefully orchestrated to study how the prairie maintained itself so well. Konza's activities include intentional burning at different intervals; reintroduction of buffalo, elk, and pronghorn; analysis of both the volume and chemical content of streams under varying conditions; and manipulation of many other factors. It is not the purpose of the research to provide information with practical application, but the increased knowledge will certainly create practical spin-offs.

As the prairies gradually became inhabited

Experiments going on at the Konza Prairie Research Natural Area in Kansas include burning the native prairie, as was done by Indians and by natural lightning. The purpose is to learn what effect this has on the complicated ecosystem that allowed native prairie to be so successful.

Grasses and legumes are often specially developed for particular areas and climatic conditions. This Bermuda grass developed in Georgia has revolutionized the quality of grazing there.

by farmers, the ground was plowed and put to crops, mostly corn. The land that remained as prairie was used as pasture and grazed heavily and differently from the prairie grazed by wild inhabitants. Possibly most importantly, frequent burning of the prairie stopped. As farming progressed, the remaining grasslands not only were pastured but were sometimes cut for hay, further changing the ecosystem.

The number of cattle that can eat from an acre of pasture and how that pasture can best be managed depends on many factors. Rainfall in the East is normally ideal for forage growth and heavy grazing. Pasture in the arid West supports far fewer cattle. In the more productive parts of the country, the same ground can be used for both pasture and hay. In the major eastern and midwestern dairy areas, for example, hay is cut from fields, but cattle graze the same fields before and after haying because forage regrows rapidly under good conditions. This multiple use is also common in irrigated areas, as we'll see when we look at ranching operations in the Colorado mountains.

In dry areas, however, many thousands of acres of grassland are pastured but are not made into hay at all. The amount of this relatively dry grazing land is huge. The total land available for crops in the United States comes to about 400 million acres, which could be increased if needed. Grazing land in the forty-eight states is hard to estimate but, including both private and federal land, is probably between 700 million and 800 million acres.

More than 30,000 ranchers and sheepherders use public grazing land, and more than 6 million animals graze on it each year. Not included in these figures are an estimated 65 million acres of private land used only for pasture. It's obvious there is a lot of land used to support our grazing animals, much of which would have no other commercial use.

Environmentalists have long argued that much of the western dry land should never be pastured or should be pastured much more lightly. Since most of this land is publicly owned, the issue is subject to political pressures that preclude definitive decisions in either direction. There have been many efforts over the past seventy or eighty years on both public and private land to improve for-

Many of the seed crops for grasses are grown in Oregon and shipped all over the world. This is crimson clover, a legume used for pasture and hay.

The most commonly grown forage today is alfalfa, a legume. In many areas of the country, it has completely replaced grasses as a hay crop. It can be pastured in the field, made into hay, or cut from the field as "green chop" and fed to animals while it is still fresh and green. In effect, the pasture is taken to the animal.

age. Unwanted brush and trees have been crushed, beaten, burned, and killed with herbicides; range has been seeded and fertilized; grazing animals have been limited to fewer numbers and moved from one pasture to another more frequently. Range management is a science that is still in its infancy.

Currently confusing the issue is a well-publicized but controversial theory that the dry land should actually be used more heavily, that the urine, manure, and hoofprints of the cattle or other users would increase vegetation, and that the trouble all along has been that we haven't been using the land enough! Because the dry southwestern rangeland is an area that grows slowly, studying it is a lifetime process, and any valid conclusion about this most recent theory can only be reached after years of range management experiments.

Beef cattle herds are usually raised with pasture as their summer home and the grass

as almost their only summer feed. Whether they are on government or private land, they can survive just on native forage for the summer. Sheep are excellent grazers and can subsist on very poor land; they usually are moved frequently so they do less damage to sensitive pasture. Dairy cattle have traditionally also used pasture for much of their summer feeding, although cows being milked can't wander far from the milking parlor, and the nutritional demands of cows being milked require supplemental grains or other high-protein feed.

For dairy cows the feeding philosophy is moving in the opposite direction from grazing. Increasingly the tendency is to bring the feed to the animals rather than take the animals to the feed. One solution has been to chop green crops such as alfalfa in the field and deliver them fresh to confined cows. In the large western dairies, in California especially, cows are usually not released to pasture at all and depend on hay, corn silage, and grain supplement being brought to them.

Beef cattle in feedlots, of course, don't graze either. Their feed is usually hay and corn silage, with grain supplements added, although a list of the varied crops and crop residue fed commercially to beef cattle would occupy an entire page.

Hay

Pasture may be satisfactory for the summer, but in much of the country it's neither available nor possible in the winter months. This is where hay plays a major role. In effect, hay is pasture stored for winter use. Hay has been cut from natural grasses for thousands of years, but only relatively recently has it been cultivated as a crop, and only in the last hundred years has its handling become heavily mechanized.

Sixty million acres of hay are grown in the United States, with a value of about $10 billion. An average ton of hay brings about $50, although hay prices are very sensitive to quality considerations as well as supply conditions. Prices go up and down rapidly. The productivity of hay land varies immensely. Hay from dry ground in the West may come

Beef cattle almost everywhere in the United States are on pasture during much of their lives. Although conditions vary greatly, most beef cattle probably gain at least half their weight from pasture. Dairy cattle, however, are being pastured less and less. The pasture is increasingly brought to them, usually in the form of hay.

a giant green loaf of bread. Cattle can be brought into the field to feed directly from the loaf, or an ingeniously designed low-slung trailer can be slid under the entire pile and trailed to the farmyard or to the alfalfa dehydrating plant. Hesston Corporation, one of the major manufacturers of farm equipment, makes a stacker that produces loaves 8 feet wide, 20 feet long, and 15 feet high, weighing as much as 6 tons.

The result of all this mechanical ingenuity is not only greatly decreased labor costs but totally different storage needs. Lifting hay into a barn mow is an activity of the past. Hay is now routinely stored outside, usually protected only by plastic or stored in inexpensive metal buildings. Stacked bales in dry country can sit outside with no protection and little loss of quality.

Forage and hay will be major crops so long as we continue to be a meat-eating country, and so long as they are major crops, American ingenuity will improve how pastures can be used and how hay can be handled.

Lesser Crops

There is a group of crops that are still major in dollars or in acreage, although they don't match the blockbusters. The blockbuster crops we have looked at all occupy 50 million to 70 million acres. After the blockbusters there is a large drop in acreage planted, to less than 17 millon acres, for the next group of crops: sorghum, barley, cotton, oats, and corn for silage. Then there is another major drop, to a range of 1 million to 3 million acres for sunflowers, rice, peanuts, dry beans, and potatoes.

When dollars instead of acres are looked at, however, the picture changes, because some crops with small acreage have a high return per acre. Down the line from the blockbusters in income is tobacco, which brings more than $3 billion to the growers each year but occupies less than a million acres of cropland. It's a really high dollar-per-acre crop, bring-

ing close to $4,000 an acre. As a result it is the sixth-largest crop in dollars, behind the blockbusters and cotton. By contrast, wheat grosses for the farmer only something over $100 an acre; corn less than $300. Some of the very small crops bring a giant return per acre. Strawberries, for example, bring about $10,000 an acre.

Don't confuse high prices with high profits, because labor and other costs may absorb much of the profit. Generally, high-return crops are also high-labor crops. It is understandable that strawberries, which take two years to produce a crop and must be carefully picked by hand, are an expensive crop to raise. Tobacco is also a high-expense crop that involves plenty of hand labor. Still, if we plan to kill ourselves by smoking, it's comforting to think we're doing it with an expensive crop!

There are many crops still considered major that don't occupy the same acreage as the blockbuster crops. Barley goes mostly for animal feed, but malting barley, used in beer making, accounts for about a fourth of the production. Barley is part of a second tier of crops that occupy 10 million to 17 million acres. Sunflowers are among the crops in the next smaller group, at 1 million to 3 million acres. Sunflowers can be grown almost anywhere, but North Dakota dominates production. Almost all sunflower seeds are crushed for their oil.

Insects

Controlling insect damage is a way of life for farmers. There are few crops that insects do not attack seriously enough to require strong counterattacks. To ignore insect damage would decimate agriculture. On the other hand, while chemists have been increasingly creative in designing new insecticides, the negative side effects of these insecticides on humans have also become better understood. Farmers and agribusiness people realize that insecticides must become safer while remaining effective. Although alternative agriculturists are convinced that insect damage can be completely controlled through "natural" means (we *are* making progress in natural control), we will for the foreseeable future be partially dependent on chemical controls. Through a sophisticated system of using "good" insects to combat the "bad" ones, and combining this technique with chemical controls, we can already manage with fewer chemicals and less harm to the biosphere. Still, we have a long way to go.

It's difficult to tell the good from the bad insect by looking. The cotton boll weevil (bottom left), the boll worm (left), and the alfalfa weevil (bottom) are all serious pests that cause immense damage. The boll weevil almost put the South out of the cotton business, and the boll worm is a problem not only in cotton but in corn, tobacco, tomatoes, beans, and other crops. The alfalfa weevil, uncontrolled, can decimate an alfalfa field.

Facing page: Insects also come in "good" varieties, and much more is being learned about their usefulness in combatting "bad" insects. Lady beetles (top right and left) are voracious eaters of "bad" insects—in the right-hand picture they are consuming aphids. Insects prey on insects (bottom left). The white shapes are pupae of the braconid wasp, which lays its eggs inside the tobacco bornworm. The honey bee (bottom right) is another "good" insect. Many crops couldn't pollinate without them and they produce more than 12 million gallons of commercial honey each year.

Sorghum (top left and right) is a grain somewhat similar to corn but adaptable to drier soils. A type different from this short-grain variety grows three times as tall and is cut green and used for silage, much as some corn is.

Sugar beets (lower left) compete with the sugar made from cane. The economics of sugar is more closely related to international politics than to agriculture, and growing sugar profitably in the United States is difficult.

Not in the high-dollars-per-acre bracket are the small grain crops: rye, oats, and barley. They are grown much as wheat is, but rye and oats bring in even fewer dollars per acre than wheat; barley's price is helped out by its use for beer making. Oats is simply one of those crops that is being phased out as a grain. Whereas we harvested 1.5 billion bushels in 1955, in 1986 we brought in only 25 percent as much. Recently we've actually begun importing oats from the Scandinavian countries. Their oats exports are heavily subsidized by their governments, bringing the price down, and ships that would otherwise return empty from our own exporting of other grains to Europe provide very cheap transportation. It's an odd world.

Malting barley, used in beer, represents about one-fourth of the barley grown, but most barley goes for animal feed. North Dakota is the largest grower, but because brewers usually contract for the growing of their specialized malting barley, it's grown in many different places convenient to the breweries. The San Luis Valley in south-central Colorado, for example, is a major supply area for barley for Coors beer.

Sorghum doesn't make it to blockbuster status but is nevertheless a giant. It's actually used mainly as a substitute for corn in animal rations, both as a grain and as silage. Sorghum grown for both grain and silage covers less than 18 million acres and in the 1985 crop year had a value of a little more than $2 billion, compared to corn's 60 million acres and value of more than $21 billion. But sorghum does better than corn under dry conditions, which is one good reason for growing it rather than corn in parts of the country. It doesn't offer the "manufacturing" possibilities of corn, but it's a useful feed. Kansas is the primary grower, followed by Texas, then Nebraska—all states with limited water supplies and big cattle populations that provide a nearby market.

Some crops like sugar beets do well under many conditions but are planted only where processing facilities are available. A processor, in fact, may come to an area and agree to build a processing plant if the area farmers will sign up a guaranteed number of acres to the needed crop.

The Southern Crops

Some large crops are restricted by their climatic needs—they grow only in the South. Among these are cotton, tobacco, peanuts, and rice. Rice not only has the need for a relatively warm climate, but also requires access to major amounts of controllable water. It isn't that the South can't produce the blockbuster crops, and to some extent it does, but there is obviously an advantage for the region to produce crops that only it can grow.

Many of the southern crops, in addition to climatic needs, had another requirement that profoundly influenced the history of the South—cheap labor. Cotton, tobacco, and rice all required great amounts of labor, and the early plantation owners solved the problem in basically the same way that present-day farmers do: they invested capital. Farmers today have been investing their capital in machinery. Plantation owners invested in slaves. It is estimated that in 1700 there were about 2,500 blacks in South Carolina; by 1765 there were almost 100,000, with twice as many blacks as whites in the state.

Slavery allowed high-labor crops to be successful, and the high-labor crops that the South prospered with were also the crops that only the South could grow. Following the Civil War the crops remained the same, and although labor costs increased somewhat, they still remained relatively low. The region's dependence on cotton decreased as the result of a number of factors, primarily soil depletion and insect problems, as well as increased foreign competition. The advent of peanuts and soybeans filled the need for alternative crops.

As field labor became less available following World War II, the pace of mechanization stepped up. Today most of the South's crops are as highly mechanized as the blockbuster crops are. The farmers have substituted machinery for people as labor has become scarcer and higher priced. There are, of course, some crops that cannot be mechanized thoroughly; tobacco is one, and many fruits and vegetables are among the others. In the border states of Arizona, Texas, and California, farmers have been able to solve the problem by importing Mexican labor while at the same time making an effort toward mechanizing.

Let's look at some of the southern crops, keeping in mind that, climatically, California is also southern.

These southern crops are dwarfed in acreage by cotton, which usually runs to about 10 million acres.

Facing page: Rice (right) is grown in water and handled unlike any other crop, with much of the work being done from airplanes. Tobacco (far right) has less than a million acres but immense value per acre. This is flue-cured tobacco in flower in Georgia.

Peanuts (bottom left and right) grow underground, but, like potatoes, their management has been completely mechanized. The mechanical digger on the right lifts the roots with the peanuts from under the ground and deposits them above ground in rows. A harvester lifts them and separates the peanuts from the roots.

Cotton

Cotton conjures up the historical view of slaves bent over, picking white bolls and dragging large sacks, as they worked their way slowly through a cotton field. I don't know of any fields that are handpicked any longer. Cotton is now a mechanized crop, with monstrous multi-row pickers churning their way through the fields, often working into the dusk of evening. Capital has replaced labor, although it has come later in cotton.

Mechanical harvesting, for example, only came into common use during the 1950s, and by 1960 it was estimated that half the crop was still harvested by hand. By 1970 hand-picking had almost entirely disappeared. Hoeing of cotton by hand was a major cost until the 1950s, when chemical weed control became possible and greatly reduced costs.

With more mechanical equipment larger farms became both possible and necessary. As recently as 1949 the average farm had only 24 acres of cotton. By 1969 the average was 58, which by 1982 had soared to 256 acres. The number of farms growing cotton dropped from 200,000 in 1969 to 38,000 in 1982.

During the era of slavery, immense fortunes were built on the relatively free labor, and cotton was almost the only crop grown in much of the deep South. Even after the demise of slavery, low labor costs allowed cotton to continue to dominate. A combination of factors broke its grip on the South. Oddly enough, one was a tiny insect, the boll weevil.

The boll weevil had such a devastating effect that a monument to the insect was

Mature, long staple cotton, ready to be picked. Cotton was almost wiped out in the South because of the boll weevil, and the damage done by the insect is credited with forcing the South to diversify into other crops.

erected in 1919 in the town square of Enterprise, Alabama. In part, the inscription reads:

> In profound appreciation of the boll weevil and what it has done as the herald of prosperity, this monument was erected by the citizens of Enterprise, Coffee County, Alabama.

What the inscription means is that because of the immense damage wrought by boll weevils, southern agriculture was forced to diversify out of cotton, and that has helped rebuild the South's economy.

The boll weevil is still a factor in cotton growing. After more than a hundred years of damage and many millions of dollars in research, not to mention millions spent each year to control the insect, the boll weevil remains a major pest of cotton. It's hardly the only pest, though, and a partial list would include cutworms, thrips, fleahoppers, aphids, tarnished plant bugs, bollworm, pink bollworm, beet armyworm, spider mites, cabbage loopers, and stink bugs. Chemicals keep them under control, but the side effects on other life, including ours, continue to be strongly debated.

Clearly we are moving toward the use of "good" insects to help fight the "bad" and toward much more sophisticated applications of fewer chemicals in most crops, including cotton. However, we have in no way done more than control insects such as the boll weevil, keeping them at a level at which their damage to the cotton does not drop the crop below a profitable level.

Cotton in the United States is grown as an annual, started anew from seed each year.

Because of the availability of labor, the mechanization of cotton harvesting took a long time. Estimates are that, as late as 1960, half of cotton was still hand picked. As machine harvesting came in, farms required more capital and less labor and grew much larger. Cotton-picking machines (left) may go through the fields several times, each time harvesting only the mature bolls.

The crop is actually a perennial that will continue to grow for many years in tropical countries without replanting, but it is more productive here as an annual.

The same standard row planter that is used for corn and soybeans is used for cotton. Its seed is sensitive to cold soil, so farmers use frequent temperature checks to determine planting time. The soil temperature needed varies with location and plant varieties. In the Rolling Plains area of Texas, for example, a soil temperature of 60 degrees at a depth of 8 inches for ten days prior to planting is the standard used.

Weed control is vital. Rotating cotton with other crops from year to year can help control weeds. Mechanically cultivating the ground is also commonly done, and when the crop gets too large to allow a cultivator

Most cotton grown in the East and in the Mississippi Delta area is American Upland, which has medium-length fibers. In the heavily irrigated areas of Arizona, cotton with much longer fibers is produced. More valuable, it still occupies a very small percentage of the total cotton acreage.

through, a period called layby, a longer-lasting herbicide may be put on to control weeds for the rest of the season.

Herbicides are routinely also put on before the cotton is planted, or after it is planted but before it emerges. The trick is to put on enough to kill the weeds but not enough to kill the cotton. The herbicides used today are very selective and can be chosen to work on the particular weeds of a particular area. They must be used carefully, and crops that follow in the rotation must also be consid-

ered because the herbicide from one year could inhibit the crop of the following year.

About 35 percent of the cotton crop is irrigated. As discussed in the chapter on water, the High Plains country of Texas is crossed north to south by the 20-inch rainfall line. Cotton grown there varies from cotton that is not irrigated at all, to cotton that usually gets 5 to 8 inches of water before planting, to full-irrigation cotton, which in the Texas High Plains means a foot to a foot and a half of water. In Arizona, which gets much less rain-

fall than the High Plains, cotton is fed as much as 60 inches of water during the season.

Cotton harvesting uses machinery completely unlike other major crop harvesters. There are two different harvesting techniques. About three-fourths of the crop is picked with a spindle-type picker. The leaves of the plant are removed with a chemical spray about two weeks before picking. The picker machines remove only the open, mature bolls; about a third of the picked crop is later picked over a second time as more bolls open.

Most cotton in Texas and Oklahoma is harvested with a stripper rather than a picker. Strippers work faster than pickers and strip the plant of all its growth in a single operation, including not only the open bolls but also the closed bolls and the useless foliage and stems; the useless material, called trash, is separated mechanically later on.

Increasingly cotton is stored in modules, large rectangular compressed stacks, to await its trip to the gin. As most schoolchildren know, the cotton gin, patented by Eli Whitney in 1794, helped make cotton the South's major crop. If you've ever tried to separate a cotton seed from its lint, or fibers, with your fingers, you can only have great respect for the invention of the cotton gin.

The present-day gin separates trash and seeds from the lint before the latter is baled. The baled cotton lint is pure white with no visible impurities, and the average bale weighs 480 pounds. The farmer pays about $45 to $50 a bale to have his cotton ginned and baled. The cottonseeds removed by the ginning process are a valuable by-product that is crushed for cottonseed oil. The cake

and meal that remain after the oil is crushed out are used for cattle feed. Cottonseed represents about 10 percent of the total value of the cotton. The cottonseed oil, of course, competes with both corn and soybeans in the edible oil market.

Most of the cotton we grow is known as American Upland, which is also the standard in most other cotton-producing countries. The lint from this ranges from ⅞ inch to about 1¼ inches in staple length—its individual fibers are mostly that length. It is called a medium-staple cotton. By contrast, about 1 percent of American production is ELS cotton, Extra Long Staple, which has a

Harvested with the seeds tightly attached to the lint, cotton has to be ginned. After being separated, the seeds are crushed for their oil, and the lint is compressed into bales that weigh about 500 pounds each.

staple length of 1⅜ to 1¾ inches. It is used in the highest-quality cotton products and thus brings the highest price, but unfortunately it also has the highest production costs.

Texas is by far the largest cotton state, growing almost half of the total production. California and Mississippi follow Texas, but seventeen states produce at least some cotton, with Missouri being the farthest north. Extra Long Staple cotton is grown mostly in west Texas, New Mexico, and Arizona.

If you drive through the cotton area of Texas or the Mississippi Delta or California, and pass field after field of white bolls stretching for mile after mile, you'll find it difficult to believe we are not the world's largest cotton producer. The truth is that China has more acreage planted to cotton than we do, and the Soviet Union produces almost as much as we do. The three nations, China, the Soviet Union, and the United States, grow about 60 percent of the world's production, although some seventy-five countries grow cotton.

Our problems with cotton exports are typical of many of our crops in the world markets. The competition can simply outsell us. They produce more cheaply, usually because of lower wages, and in both the Soviet Union and China, prices are completely controlled by the government. In 1979 we exported 9.2 million bales, but by 1985–1986 our exports had been cut to 2 million bales. Not to worry. By subsidizing our own exports with federal money, we have brought our export price down toward world levels and are again beginning to be competitive. The cotton farmer is getting paid a livable price, the foreign buyer is getting a competitive price, and Uncle Sam is keeping everyone happy. It is working, our exports of cotton are increasing substantially, but the taxpayer is footing the bill for it.

The acreage we devote to cotton has changed immensely over the years we've been producing it. At the end of the Civil War, acreage was down to less than 8 million acres. By the mid–1920s it had grown to 44 million acres, but by 1966 it had dwindled back to 10 million acres, which was also the total in 1985. Acreage fluctuates widely from year to year now, depending on price, exports, and the way government subsidies are handled. While cotton may eventually become economically impossible to produce, the alternative crops may also become less attractive. The situation is one we've encountered before.

Tobacco

A maturing tobacco field in the dampness of early morning mist is to me a beautiful place to be. The gracefully shaped tobacco leaves are a lovely soft green color, and the smell of ripening tobacco, heavy in the field, is both acrid and sweet. The warmth of the morning sun brings out the scent. In some ways it is even more beautiful than a ripe wheat field.

The giant leaves are a far cry from the fine seed that is planted in meticulously prepared beds. In earlier days, these starting beds were moved every year so that disease from previous crops couldn't attack the new crop. They still must be separated by considerable distance from the curing barns to prevent diseases moving from the remains of the old tobacco to the tender seedlings. In some tobacco areas, such as Lancaster County in Pennsylvania, the tobacco seedbeds have traditionally been "steamed"—the ground sterilized each year with several hours of steam applied to it. Today chemicals mostly do the job of protecting the tender young plants from disease and from weed competition. Plastic covering frequently is used on the seedbeds to soften springtime climatic variations. Tobacco seedlings begin life as very tiny plants and they need all the care they can get.

The seeds are so tiny that an ounce of them, a little more than a tablespoon, covers a planting area 36 by 100 feet and produces about 25,000 plants, enough to plant 4 acres. The seeds are so small they are usually mixed with an inert material such as sand or bone meal to add bulk so they can be broadcast more evenly. It takes about two months from the time the seeds are planted until the 6- to 9-inch plants are ready to be transplanted.

Although transplanting is mostly done

Tobacco comes in many different types. The field of tobacco in North Carolina (left) grown to be flue-cured will be processed entirely differently from the air-cured tobacco grown in Pennsylvania (right).

Barns

They really don't make barns the way they used to, but then they don't need to. In earlier times barns had two functions: to protect animals and to store crops. Today both functions generally can be achieved more effectively by using structures other than barns. Just as urban dwellers have increasingly gone to single-story houses—called, oddly enough, ranch-style houses—farmers have gone to single-story storage and animal housing. The materials used for constructing barns have changed from wood and logs to aluminum, steel, and concrete, with less cost but with less romance, too.

New England barns (left) were massive structures almost always used to store hay for the long winters and to house dairy cattle. Frequently, because of the cold weather, the barns were actually attached to the farmhouses.

Midwestern barns (top right) were not normally used to house cattle—the beef cattle wintered outside. However, they did have huge hay storage areas for loose hay.

The most interesting barns are the few remaining round ones (far left, bottom right). Occasionally seen everywhere in the United States, they were most common in New England. Some, like the stone barn (bottom right) at the Hancock Shaker Village in Massachusetts, are being carefully preserved for their historical value.

although air curing is also used for cigar tobacco. Burley shows up in cigarette blends, pipe tobacco, and chewing tobacco. Most of the burley is grown in the bluegrass country of Kentucky and in Tennessee.

The production of burley is catching up with flue-cured tobacco as the world moves toward blended tobacco for its cigarettes. Currently burley's production is slightly more than two-thirds that of flue-cured, and the two together represent more than 90 percent of the tobacco grown.

Tobacco is not entirely a southern crop, although most of us think of it as such. Filler tobacco for cigars comes from Pennsylvania, Ohio, and Puerto Rico. Wisconsin's production is used mostly in chewing tobacco. Connecticut's tobacco crop is used as cigar binder and wrapper, although its production has been dwindling, partly the result of decreasing demand.

Tobacco is hemmed in by government controls aimed at keeping production down but income up. Tobacco growers in most areas are limited to a specific number of pounds of production, agreed to before the crop is planted. The number of acres to be grown is also allotted, although the grower often doesn't use all of his allotted acres, few as they are, because he can grow as much as he is allowed on fewer acres, thanks to increased productivity. Government programs that deal in strict poundage limits are currently restricted to peanuts and tobacco. On other crops the number of acres that can be planted is the controlling factor.

The tobacco programs have been highly controversial. Besides controlling production, the government decrees a tobacco support price each year, as it does for many other crops. If the tobacco fails to bring the support price when it is sold at auction, the grower is lent the money the support price would have brought, usually through his cooperative but with the loan guaranteed by the government. Eventually the market price goes up—tobacco can be stored for years—and the tobacco is sold for the grower, who repays the government for his loan, plus interest. Inflation works well with tobacco, because it keeps the price rising, getting the tobacco marketed and getting the government loans paid back.

The program is an inexpensive one as farm programs go. In fact, it has been the law since 1982 that the program cannot cost the taxpayers anything. Whether this actually has been the case is certainly debatable. It is also questionable whether the government should be in the business of controlling the production and supporting the price of a crop that it is also trying to get us to stop using. Peculiarly, one of the arguments for tobacco price supports is that without them the price of tobacco would go down, making increased use more attractive!

Rice

Growing rice has almost as much in common with raising catfish as it has with growing corn. It is unlike any other American crop, for it is partially grown under water. Rice had its American beginnings in the marshes of South Carolina and Georgia, areas wetted down with fresh tidal water from the rivers. Nothing else could grow there, but rice did wonderfully. In 1700 the crop was so good that colonists were able to export 300 tons of rice to England.

During that period rice had very high labor requirements, as it still has in some parts of the world, and the South used slave labor to fill the need. The rice industry on the East coast was wiped out by the Civil War, however, and moved to the Gulf Coast area following the war, when returning soldiers were given land there. Production is now centered in Arkansas, which produces almost twice as

much as next-ranking California. Surprisingly, though, it wasn't until 1902 that Arkansas began to produce rice.

In many Asian countries rice production still is mostly done with hand labor. The rice seeds are planted in special beds and then removed as young plants after about a month and laboriously planted in water-soaked fields one plant at a time. There are still some countries of the world where it takes 300 man-hours of labor to grow an acre of rice. In the United States that time is down to 7 hours.

Supposedly rice growing did not become mechanized in the United States until a wheat farmer from Iowa went to the Gulf Coast in 1884 and applied his wheat-growing knowledge to rice production. The rice fields in the United States have controllable water. Through the use of levees and pumps, the fields can be dried out or flooded as needed. Rice can be planted on dry ground with a regular grain drill, like wheat; the water can then be introduced after the planting. Or the seed can be dropped onto a wet field from an airplane and allowed to settle to the ground through the standing water. It seems to make little difference to the rice; however it gets planted, it grows. As rice sprouts, it does not need to be too wet, but as it grows, the water level in the paddy is raised. Weeds that cannot live in water are killed off as the water is raised. Other weeds are controlled through herbicides.

From planting time until harvest, the field operations are done by airplane. Herbicides to kill weeds may be put on as part of the ground preparation before planting or flown on after planting. Fertilizing the growing crop is done from the air, and when insects become a problem, they are attacked with spray from the air. Most estimates of rice-growing costs include about eight different aerial operations. The early morning calm around the rice center of Stuttgart, Arkansas, is frequently broken by the sound of spray planes buzzing like a bunch of angry hornets.

Before the crop can be harvested, the fields must be dried out for several weeks, at least enough so that specially adapted combines and grain carts with oversized tires can work without getting stuck. Watching these giants roll up over the water-control levees in the fields is a sight to remember.

In the rice-growing areas of the Texas Gulf Coast, it is common to "ratoon" the rice. After the crop is combined, the roots and other parts of the rice plants remaining in the fields are allowed to regrow, and a second crop is harvested. It usually amounts to about 15 to 20 percent of the volume of the first crop.

Rice isn't just rice. There are some 40,000 different varieties grown in the world. In the United States the varieties have been divided into long, medium, and short grain. Long-grain rice cooks dry, fluffy, and separate; medium- and short-grain rices cook moist and have a greater tendency to cling together. About three-fourths of our crop is long grain, 21 percent is medium, and less than 5 percent is short grain. Arkansas grows most of the long grain, and California is the largest grower of both medium- and short-grain rice.

Regular-milled white rice, usually called white rice or polished rice, is the form most familiar to the consumer. Parboiled, or "converted," rice is treated with steam pressure before it is milled to harden the grain, which helps make the rice fluffier when cooked by the consumer. Precooked, or "instant," rice is milled, then completely cooked and dried. It speeds up preparation by the consumer. Brown rice is the least processed form of rice. The outer hull is removed, but the bran layers are retained and give the rice its tan color and nutty flavor.

As for wild rice, it isn't really related to other rices, except that it is also a grass. It is native to the Midwest and for many years was harvested by hand by the Ojibway Indians of Minnesota and sold at high prices because of its limited production. Around 1972 California began to experiment with growing it as a commercial crop, and by 1980 the area near Marysville was producing about 10,000 acres of wild rice.

As I watched it being combined, I realized that, as a new crop, it hadn't yet reached a

Many of the jobs in rice growing are done from airplanes, since working the fields when they are largely under water is very difficult. This plane is planting rice; the man in the foreground serves as a guide so the rice will evenly cover the field. To manage water flow, the level of rice fields has to be very carefully controlled so that the field can be flooded and dried as needed during the growth cycle.

A mature rice plant is one of the most beautiful crops. For many years rice growing and harvesting was a very labor-demanding crop, as it still is in many parts of the world. Today, however, its production in the United States is as mechanized as most other crops, and it is harvested with the same sort of combine used for wheat and corn—with the addition of oversized tires to get through the wet fields.

standardized stage. The height of the plants varied immensely, from 3 to 7 feet, and the amount of grain in the heads varied greatly also. This is a plant just being developed. Plant breeders will take care of that, as they have corn and soybeans, and in the meantime production is high enough to make the crop profitable, since the market still considers it a gourmet item and is willing to pay for it as such. As the crop is developed and production grows, the price will decrease.

Wild rice may have a high price and a limited market, but ordinary rice is one of the world's great staples. Rice and wheat are the most widely produced crops. The average annual per capita consumption of rice for the entire world is 150 pounds, and in many countries it comes to almost a pound a day per person. In the United States each of us consumes only a little more than 50 pounds annually.

It has been estimated that more than half the world's rice is consumed within 8 miles of where it is grown. Considering its immense worldwide production, surprisingly little rice appears in world trade. American rice amounts to less than 2 percent of the world's production, but we account for 15 to 20 percent of the international trade. While our total production is only 4 percent of the amount grown in China, we export four times as much as China. As with other crops, however, China is the country to watch, for their entire agricultural production is zooming upward, and as their exports expand, they will increasingly be our competitors.

Animal agriculture may be a rather touchy subject for the less than 3 percent of us who refuse to eat animals. For the other 97 percent of us, however, eating animals is a way of life. Many of the decisions we make concerning what animals we eat, what parts of the animals and in what form, are the result of advertising, competition, and dozens of other influences. Our decisions are rapidly translated through the marketplace into changes that affect the farmer.

Beef has long been a symbol of affluence. There still remains a certain cachet to ordering a filet mignon or a T-bone, not to mention a rib roast for the family. Times and demand do change, however. The move for several years has been away from red meat toward white, from fat toward lean. Between 1970 and 1984, lamb and mutton consumption per capita dropped 47 percent, pork dropped 10 percent, and beef dropped 6 percent, while broiler chickens rose 43 percent. Economists expect that chicken consumption will overtake beef very soon.

Americans consume prodigious amounts of meat. We still eat nearly 80 pounds of beef per person each year and almost 60 pounds of chicken. Our consumption of pork remains steady at around 60 pounds. Lamb's lack of popularity has been a disaster for sheep growers, although they argue that consumption has stabilized and will possibly be on the rise over the next few years. We eat about 1.3 pounds of lamb per person, less than half of what we ate twenty years ago.

The fish we eat is increasingly homegrown rather than caught in the wild of the sea. Catfish production increased more than four times from 1980 to 1987, to about 250 million pounds, partly because of its use in fast-food restaurants. Mississippi grows more than 80 percent of the catfish crop, which is carefully nurtured in ponds. Commercially raised fish still represent less than 10 percent of our total fish consumption, but it is a rapidly growing segment of our meat supply. Whether fish farming can be called agriculture depends on your definition, but it meets mine.

The biggest single goal in animal production has always been to lower the feed ratio. Here the effort has been to get more pounds on the bird or the cow or the hog or the fish using fewer pounds of feed, or fewer dollars of feed, and a shorter amount of time. Chickens currently require about 2 pounds of feed for each pound of weight gain. Fish can do even better. Beef cattle don't come near, although cattle can eat cheaper feed because they can handle roughage such as hay, corn

Fish farming, although still a miniscule part of animal agriculture, is growing rapidly. Fish are excellent converters of feed to meat—the ratio of the amount of feed they eat to the amount of meat they produce is better than any other animal. Here a crop of catfish is being harvested from a pond in Mississippi, the center of the catfish industry.

silage, or green grass. Hogs have literally been redesigned. Longer and skinnier, they have far less fat and more meat. Beef breeds of cattle also are moving toward more meat and less fat, which means they are given more grass feed and less corn "finish" (which produces the fat marbling we used to demand but now shy away from).

Vegetarians have long complained that we would have far more food available if we didn't feed it to animals. The usual ratio given is 5 or 10 to 1; that is, we could have available 5 to 10 times the food value through eating the grain ourselves, rather than feeding it to the animals we eat. Although there may be some merit to this argument, the truth is we Americans love meat and presently don't need the additional grain that would be available if all of us were vegetarians. In addition, vegetarians seldom take into account the immense acreage of grazing land that would be of no use if it didn't have cattle on it. Finally, a sudden shift away from meat would be an agricultural disaster with which the United States couldn't begin to cope.

There is an increasingly important problem related to meat production: manure disposal. As the number of animals has increased, and the space in which they are raised has decreased, the problem of manure disposal has become more critical. This is particularly true in the more populated areas of the country. Ranchers in the Rockies who have beef cattle on pasture encounter little problem with manure disposal. It generally takes care of itself exactly where it should go—on the pastures. At the other extreme, however, chicken growers in Lancaster County in Pennsylvania, along with the hog, dairy, and beef finishers there, have generated a major problem that they are still trying to solve.

Just imagine how much manure there is to dispose of. Although quantities vary, a single dairy cow eliminates more than 25,000 pounds of excrement a year. If a farmer is milking, say, 50 cows, he has to dispose of 625 tons of manure; a 3,000-head dairy operation boggles the mind! How this manure is handled has become a specialty of agricultural study—and dispute.

Increasingly, manure is handled as a liquid slurry rather than as a solid because liquids

can be easily moved by pumping. Dairy and steer manure on medium-sized farm operations is likely to be mixed with water and put into huge open storage tanks where it is mechanically stirred and decomposes somewhat. Then it is taken to the fields in tank trucks that spray it onto the soil surface or inject it into the soil. In general, urine has more nutrients than solid matter, and slurry usually does a better job of retaining the urine than other handling techniques. Slurry also provides a convenient way of storing manure until it can be put onto the fields, particularly through the winter months when manure put onto frozen fields only washes away into the water supply.

Although manure is valuable, its chemical value is not great when reduced to actual figures. A corn farmer, for example, often uses a mixed chemical fertilizer that is 20 percent nitrogen. If he wants to put 100 pounds of nitrogen on his field, he has to handle only 500 pounds of chemical fertilizer. On the other hand, cow manure is likely to have only 10 pounds of nitrogen to a ton, which means that the farmer has to handle 10 *tons* of manure to get 100 pounds of nitrogen. Moving that much material is time-consuming and costly.

That isn't, of course, the end of the argument for manure. The organic matter—the physical bulk—of the manure increases the

Almost all chickens spend their lives indoors, and most hogs are now also raised in confinement housing. Dairy cattle are partly kept outdoors, mostly depending on the climate, but are milked indoors in milking parlors. Beef cattle, except in a small number of feeding situations, spend their lives outdoors, either "on range" or in feedlots. Sheep live almost entirely outdoors.

In the Rockies and much of the West, cattle and sheep graze on pasture, some of which is privately owned, some owned by the government. The land is seldom managed, except for limiting the number of animals on it and the amount of time they graze it. Some private pasture is irrigated. Dairy cattle, on the other hand, are almost always on pasture that has been improved over the natural grass.

handling quality of the soil (the "tilth") when worked into the soil, allowing it to build up useful bacterial content, hold more water, and erode less. The organic matter of manure can also be improved by composting before it is worked into the ground. The manure and other organic materials are processed in any of various ways that involve stacking them and letting the heat that is generated help bacteria break down the materials.

Some proponents of alternative agriculture maintain that manure is wasted. However, I know of no manure that isn't put onto the soil. It certainly may not be added to the soil in the safest or most effective way, but it is deposited there, for there aren't many other places for it to go. It is useful for the soil, but its handling must be improved to limit the problem of contaminating water. I don't believe that manure is the answer to everything the soil needs, nor that productive farming can be done using only manure as fertilizer.

On the other hand, it must be disposed of and it should be put to positive use. Before any really satisfactory solution to the manure problem is arrived at, some animal farming is going to be squeezed out of increasingly urban areas because of manure disposal difficulties. The pollution of water by inadequate manure disposal, plus odor and potential disease problems in urban neighborhoods bordering farmlands, will be a source of conflict for many years.

How does meat get grown and to our supermarkets and restaurants? For the answers, we'll go to the meat producers: the mountain ranchers who move their cattle into summer pasture in the high mountains; the chicken growers who raise chickens in near darkness in contemporary houses; the hog farmers who are rapidly changing to raise their animals more as chickens are raised; and the sheep producers who cling to the old ways, though not very profitably.

Beef Cattle

The beef cattle business is divided into a number of segments, and as in the rest of agriculture, the goals of one part are often at odds with those of another. The ranchers who breed the cows and raise their calves want to get the last nickel from each calf their cows drop; the feeders who buy the calves and "finish" them to market weight want to pay the lowest price for those calves. The rancher is worried about the price of hay and the cost of grazing rights on government land. The feeder worries about the price of corn and how to take the calves he buys from the ranchers to a smoothly finished 1,100 pounds, with a conformation that will grade out U.S.D.A. Choice and have more red meat and less fat. The packer worries about whether he is preparing an animal the consumer will buy and whether he's doing it at less cost than his competitors.

Everyone along the chain worries about whether the industry is raising the right kind of cattle. The demands of the consumer are changing, and the effects are felt all along the chain, right down to the ranchers, and carried by them to the breed associations responsible for the kind of cattle that are being developed.

Although the National Pedigreed Livestock Council lists seventeen beef breeds, there are actually dozens more. Basic breeds are being crossed and new breeds created, usually complete with their own breed associations, breed shows, and publicity. Each new breed is touted as the answer to all of beef's problems. Sometimes they last, sometimes they don't. It is actual production that counts, and statistics tell the story. There's not any perfect beef breed, but the knowledge of what each breed can do is increasing constantly.

The Angus breed comprised more than 20 percent of the total registries in 1985, with Herefords second. Polled Herefords, a separate breed that has been bred to reproduce without horns, are fourth in numbers. Considered together, the two Hereford breeds have the most registries. New combinations are constantly being developed, however, and existing new breeds are brought in from around the world, to be added to the registries.

Beefmasters are an example of a relatively new registered breed. They are about one-half Brahman, one-fourth Hereford, one-fourth Shorthorn. The lines are established, however, and so they are bred to each other as Beefmasters and are promoted as such. The famous King Ranch, in Texas, developed one of the first crossed-breed cattle, the Santa Gertrudis. They are about five-eighths Shorthorn and three-eighths Brahman. They were bred using Brahman to get a more dependable warm-climate animal; the Brahman are a breed noted for heat and insect resistance in India.

Some of the European breeds, such as the Simmental from Switzerland, are gaining in popularity. Charolais came to the United States from Mexico but are French in origin. Senepols, a hot-weather animal from the Caribbean, are just being developed; there are only about 1,500 registered in the entire world.

Cattle Ranching

The best place to start our discussion of the beef business is a cattle ranch. In order to discuss ranching, we first need to get a few terms straight. Heifers are female cattle that haven't yet had a calf; first-calf heifers are bred and are going to have their first calf. Cows are female cattle that have had calves. Bulls are male cattle, and steers are castrated bulls.

Colorado's Ohio Creek Valley is ranching country, the stuff of western novels. On the south side of the road that runs up through the flat valley stretches a narrow plain formed by Ohio Creek. Beyond the plain is a spectacular mountain range where Castle Rocks juts against the sky 40 miles away. To the north is the range that separates Ohio Creek from a valley that is home to the Crested Butte ski area. To the west is a stretched-out, no-nonsense mountain range, blanketed with snow much of the year, that guards an entrance to the West Elk Wilderness.

This is some of the finest ranching country in the United States, complete with cowboys who wear bandannas around their necks and like riding horseback a lot better than walking. The cattle are Hereford mostly, with a scattering of purebred Longhorns and increasing numbers of crossbreds and newer breeds. Cattle do well here, thanks to an irrigation system that diverts water from the Ohio Creek to produce lush hay—a mixture of brome, meadow foxtail, and some native timothy.

In May and June the meadows are so heavy with water they glisten against the sun. Many of the ranch houses have high-pitched metal roofs to dump the heavy snow that starts early in the fall and ends late in the spring. It is runoff from the snow that brings on the flooded meadows. Gunnison County is often the coldest place in the United States.

Cows and calves usually spend the summer almost unattended on mountain pastures. Government pasture is rented to ranchers for a small fee per head of cattle per day, with the National Forest Service (or in some cases the Bureau of Land Management) collecting the fee and determining the number of cattle each pasture area can hold and the number of days it can be used.

Gordon Headlee ranches here. Not that he's a native. He came from Texas in 1973.

"The University of Texas decided to build a campus in Odessa," he explains, his soft accent reminiscent of Texas, not the Colorado mountains. "They bought land between our place and town, and all of a sudden we had the city right at our fenceline. We had to decide either to get into the real estate business or to move the ranch. My brother stayed there and did great and richly. But I'd never played a game of golf in my life, and I didn't care anything about belonging to the country club, so Joan and I bought this place and moved three loads of cattle up here. They were Angus cows crossed with Charolais bulls."

Since then he has adapted to demand and to climate, and his cattle are now essentially Simmentals. "We just kind of eased into Simmentals," he says. "Now we're tryin' Saler bulls on the Simmental cows. One thing we want is smaller calves at birth. I'll bet those Saler crosses don't give us calves that average more than 70 pounds, so we can breed heifers more easily. First-calf heifers have trouble

Beef Breeds

The number of breeds of beef cattle is constantly increasing, in distinct contrast to dairy cattle, which are dominated by a single breed, the Holstein. Beef cattle are moving in the opposite direction, not only to more breeds, but to more crossing of breeds as well. Breed preferences are changing, as the demand for less fat in beef increases, and there are always new exotics that can be imported. The commercial cattle producer has largely responded to these changes by using cross-bred cattle likely to meet the market demand, and improved scientific measurement has helped tell him whether he is going the right direction. The gentleman farmer who wants something different and is uninvolved with market requirements has imported beef animals from around the world, and even resurrected a few native cattle. Returning in relatively large numbers after years of being an oddity, the Longhorn may well fit into our need for less fat on our beef.

Texas Longhorns (top left) were once the predominant cattle of the West, the stuff of Zane Grey novels, but with the increased interest in fat beef, the tall, stringy Longhorn lost out to English breeds. The Hereford (bottom far left) was one of the earliest of the English breeds brought to this country, but not all imported breeds have been successful. The Scotch Highland (bottom near left) has never been successful here and remains an interesting curiosity. The Belted Galloway (middle left) also has never become a major breed.

Many cattle breeds have been imported because of specific qualities of the animal. For example, the Brahman (top left) was brought to the United States from India because it can withstand heat and has high resistance to insects. It has now been crossed with many other breeds to form new breeds. When a Brahman is crossed with an Angus, the new breed is called Brangus. Brahman blood dominates cattle across the southern United States.

Some breeds long established in the United States have been developed into new breeds simply by selection. Polled Herefords (top right) are an example. The Hereford normally has horns, which can be a disadvantage in the confines of a feedlot or during trucking. Through selective breeding, the horns were eliminated, and as far back as 1901, the Polled Hereford was a separately listed breed.

Most of our beef cattle today are crossbreeds (bottom left). The cow here is a cross between Hereford and Angus and was bred to a Charolais bull, producing the light-colored calf.

The cattle in feedlots (bottom right) are usually crossbreeds. These are a cross between Herefords and Angus.

with big calves. We're countin' on those calves havin' a lot of potential for growth. We'll have to wait and see; we bred 125 cows to the Salers this year."

Gordon breeds his cows—he runs about 900 head, including yearlings—so they drop their calves in May, later than most ranchers. The lingering snow and cold in the Ohio Creek Valley make late calving advisable. The calving is done near the ranch buildings so the cowboys can keep close watch on the cows to help with any birthing problems. Newborn calves seem helpless for a few minutes, but are up and running very quickly after birth.

As soon as possible after calving is finished, the cows and calves are prepared for summer pasture. Bull calves are castrated. The Headlee Ranch uses the "rubber" method. Cowboys catch the calves and throw them, then attach a heavy rubber band to the scrotum tightly enough to stop blood circulation to the testicles. In about two weeks the testicles and scrotum drop off. Heifer calves not to be kept for breeding are spayed. Gordon uses a simple tool that requires a lot of know-how but no outside incision on the calf.

During branding, the cows and calves are brought from the field into corrals and separated. The cows are herded into chutes where they can be inoculated against various diseases and then are moved on into a dipping vat that rids them of ticks and lice. The calves are treated separately and branded in a chute where they are held tight and turned on their sides. Any calves that show signs of having horns are "dehorned" with a small round hot iron that is pushed into the site of the horn.

While being branded, the calves are also vaccinated.

The smoky branding and dehorning smells and looks pretty gruesome, but after the calves are set upright and released, they jump out of the chute and look around, bawling for their mothers. They pay no attention to the brand that has singed the hair off their skin.

Branding is necessary because on the first of June the cows and calves are trucked to government-owned grazing land for the summer. Other ranchers' cattle may be on adjacent allotments, and in the fall when the cattle are rounded up they have to be separated by owner. The brand identifies them.

Before the cows and calves are trucked to the allotment land, they are separated and put into different trucks because the calves, which weigh an average of less than 200 pounds, are liable to be trampled by the cows, which weigh more like 1,000 pounds. Unloaded after the trip, each cow finds its own calf—a remarkable feat.

Because the climate of the Ohio Creek Valley makes late calving desirable, Gordon's cows have to be bred in late July and August. They are already up on government land for summer pasture when the bulls are turned in with them. Cows have a gestation period of a few more days than humans.

The cattle on allotment land are checked regularly, although none of the Headlee ranch hands stay with them. The water supply is frequently monitored because water holes may go dry. There are other risks with the cattle on mountain ground, 20 miles from the protection of the ranch. Some risks can be minimized, some just remain a risk. There are cattle thieves and an occasional bear or mountain lion. There are damned fools with rifles and bad eyesight. There is larkspur, a plant that is poisonous to cattle. And sometimes cattle just get sick. But after a summer of grazing, the Headlee Ranch has usually lost no more than one or two of their cattle.

Grazing on Government Land

The political debate about the use of government land for private grazing is one that may become more heated in future years. Western ranches, when they are bought or sold, are frequently advertised as "4,000 deeded acres,

10,000 acres lease." The government lease is considered part of the ranch. It adds value to the ranch in spite of the fact that the lease can be canceled by the Forest Service or the Bureau of Land Management without any legal difficulty.

When I asked Gordon Headlee about this, he looked at me rather sharply. "Well, naturally the lease value is built into the ranch price," he said. "If you asked any of the government people the circumstances under which a lease might be canceled, they'd agree it would have to be a pretty extreme situation. At least half of them would say they'd never heard of a cancellation."

Leases can be bought and sold separately

The cows and calves are remarkably able to fend for themselves during the summer. The calves rapidly gain weight nursing from their cows and gradually begin to graze as well. The cows maintain themselves only on the grass they graze. Normally when they are brought out of summer pasture in late September or October, the calves are separated from the cows and weaned. The cows have been rebred before going into the mountains.

pasture. They still have another 400 or 500 pounds to gain before going into a feedlot for "finishing." The decision concerning their future depends on a complex set of factors. As is so common in growing crops or animals, there are choices to be made, and how those choices are made may determine whether the farmer or rancher survives or disappears.

If it's been a good year for hay on the ranch or if hay is cheap somewhere nearby, the rancher may decide he'll keep the weaned calves. Even if hay is cheap, however, if the market for calves is high, he may decide it is better to sell them off. The decision depends also on whether it involves heifer or steer calves. Gordon Headlee has already decided, before time to spay the heifer calves in the spring, how many heifers he wants to keep for replacement purposes and how many will be spayed and eventually go on to market. He may also keep a few bull calves for breeding purposes.

He usually keeps some of the calves on pasture and hay through the next year, then sells them as yearlings. There are several ways to sell them. Most recently, they have been sold through a television auction. Ranchers have discovered that animals in a beautiful mountain setting sell better than animals in a corral. A video photographer comes to the ranch and cowhands move the animals past the camera. Buyers from feedlots fly into Denver, sit comfortably in a hotel lobby, and bid on the cattle that appear before them on a television screen. The buyers are provided with statistics on the number of animals, approximate weight, and owner. It's a lot easier than traipsing around hundreds of miles of mountains to find each herd.

Fewer cattle are being finished on small farm or ranch feedlots (left); more are going into giant feedlots (right). The sizes of feeding operations vary greatly and are changing rapidly. In Texas there are fewer than a thousand feedlots that have less than a thousand head; Nebraska has almost ten times that many. In Nebraska, however, there are only five lots with more than 32,000 head, whereas Texas has thirty-five. During the past twenty years, Texas has tripled the number of its large feedlots, and the number of cattle fed in the Great Lakes and midwestern states has dropped greatly. Kansas and Nebraska have held their own with small and medium-sized lots.

Finishing and Packing

Eventually, at a weight usually somewhere between 700 and 900 pounds, the cattle go "onto feed." Feedlots are often called, somewhat disparagingly, "factory farms." There is some truth in the name; the cattle are kept in pens, usually about 200 to a pen, with nothing to do but eat, produce manure, and get fat.

Their diet is totally controlled. In big feedlots computers figure a ration for each pen, determined by age, weight, breed, sex, time of year, and a handful of other factors. Trucks move from the feed chutes to the feed troughs, dump their loads, and roar back for the next load. It's, well, a factory operation, even though it still requires a lot of personal judgment about handling, rations, and market weight. In the end, it's a scientific attempt to get the cattle to the weight and condition that the packing house wants.

Call them factory farms or whatever, feedlots are big business. With 20,000 cattle weighing 800 pounds, bought at, say, sixty-five cents a pound, a feedlot has more than $10 million invested, and that's before feed or labor costs. Not a large feedlot but, in anyone's calculations, a big business.

The feedlot business has been evolving over the past fifty years. Back in the 1960s it started moving from the Midwest and from small "family-size" feeders to giant corporate

The purpose of feedlots (top left) is to get feed to cattle as conveniently as possible, with minimum exercise on the part of the cattle. The feed that goes to each pen of cattle is controlled by sophisticated computers and electronic equipment (bottom left). Usually feed is mixed and then dumped precisely by weight into trucks that take it to individual pens. Feedlots know the total amount and kind of feed consumed by each pen and thus the cost of feed for each pound of gain.

Along with feeding, feedlots oversee the health of their animals, medicating the animals (right) either routinely or to counter specific illnesses. Medication and its potential for carry-over into the meat have long been a subject of controversy in the meat industry.

Many cattle are "hedged" to minimize risk, but feeding is still a chancy business. If, for example, you own a thousand head of cattle on feed, and the price on the day you sell drops by 2 cents a pound, you've lost $22,000. Of course, if the price goes up 2 cents, you've made an extra $22,000.

feedlots. Texas has more than tripled feedlot production in the past twenty years. Kansas and Nebraska have also done well or held their own, but the Corn Belt and Great Lakes states have dropped more than 20 percent.

The growth in Texas has been mostly through huge feedlot operations. In Texas today there are 35 feedlots that have more than 32,000 head each, only 10 that feed between 1,000 and 2,000 head, and 852 feedlots that feed less than a thousand head.

However, in the number-two state, Nebraska, the figures are very different. Family-size feeding is still very important. In Nebraska there are only 5 feedlots with more than 32,000 head and 180 with between 1,000 and 2,000 head. There are 9,270 feeders that handle less than a thousand head—more than ten times as many as in Texas. Nebraska has strong "anti-corporate" farming legislation, although it is too new to have had much effect in the feedlot business. Smaller feedlots, certainly those under a thousand head, are much more likely also to be in the farming business, growing their own hay and grain. The big Texas feedlots do have farmers raise feed for them on contract, but for the most part they simply buy "off the market" wherever they can get the quality they want at the price they want.

As in so much of agriculture, the possibility of substitution allows for a lot of jockeying in the price of feed. Several dozen cattle feeds are available, many only on a localized basis. They include the standards, of course, of corn, sorghum, hay, and silage of various kinds, as well as such exotics as avocado seed meal, dried bakery products, beet top silage, steamed bone meal, coffee grounds, cotton

Branding Time

There is still romance left in agriculture, and branding time on a cattle ranch is one of the remaining symbols. More than almost any farm chore, cattle handling has resisted mechanization, and separating cows and calves in the corrals during spring roundup is still strictly a question of manpower. The purpose of branding is to get cows and calves ready for a summer on range, checking their health and branding them so they can be identified in the fall roundup. Branding time becomes something of a social gathering, because it requires a lot of help.

While the primary purpose is to brand calves so they can be identified, branding time also gives the rancher an opportunity to inoculate both cows and calves against a number of diseases. After branding, the calves seem little affected and immediately find their mothers.

Facing page: Branding involves temporarily separating the cows from their calves. After they are reunited, they are usually given several days so that stress can diminish. Then they are separated again to prevent the calves being trampled, and loaded into trucks separately (top right) to be trucked to the summer pasture. Unloaded from the trucks, the calves are immediately able to find their mothers from among hundreds of cows.

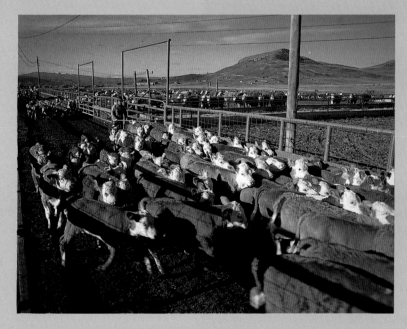

In most states in the West, whenever a beef animal is bought or sold, a brand inspector checks the brand to see that it matches the animal's description. Cattle rustling is almost, but not quite, a thing of the past.

gin trash, hydrolyzed feather meal (very high in protein), dried citrus pulp, spent hops, sugar cane bagasse, and on and on.

From the feedlot, the next stop for cattle is the final one. The average animal going to the meatpacking plant weighs 1,100 pounds—steers a little heavier, heifers slightly lighter. With increased emphasis on leaner meat, this weight is likely to decrease.

The packing business has changed from the days of Carl Sandburg's "Chicago, hog butcher of the world." It is concentrated among very few companies, most of them subsidiaries of giant conglomerates. The small packers are mostly gone, both because of the costs of having federal inspection and because of the economies of size. Today's plants, built mostly on the edges of small cities near the feedlots, are all stainless steel and automation. No question about their efficiency.

There is question of their numbers, however. Feedlot operators scream because competition among buyers has all but vanished.

Ranchers still usually have their cattle bid on by a good many feedlots, but feedlots often have their cattle bid on by only one buyer. It's a problem, and its solution isn't very clear.

Though few in number, the packers are immensely competitive, and this competition gives the consumer a chance to be heard. The supermarkets that actually retail the beef have a hotline to the packers, who have a hotline to the feedlots, who make their wants known pretty quickly and loudly to the ranchers. If consumers really know what they want in beef, they can create change.

The USDA Choice grade of beef carries with it 28 to 30 percent fat; Canadian finishers have for a long time produced beef that has 20 to 22 percent fat. In their hurry to get leaner beef to their customers, U.S. packers are trimming the outside fat from the product, but until the breeding and feeding industry changes, the marbling will be the same. Marbling produces beef that has a taste we're accustomed to and a tenderness that we seem to want.

Beef may indeed continue to lose out to chicken, but it won't be for lack of trying to stem the losses. The packer is decreasing fat by trimming. The feedlot operator is making an effort to cut back on feeding, both in terms of time and final weight, to reduce fat. The rancher is changing his breed mix, aiming away from short, stocky cattle toward taller, leaner animals the feedlot can finish leaner. Can beef catch up with its leaner competition? We'll see. No question who will decide: the consumer.

During its life, an animal may be moved many times. A calf goes from the ranch to summer range, back to the ranch, to a sale yard, possibly to winter wheat grazing, to a feedlot, to a packing house, and finally to a retail store. Trucking is one of the major agribusinesses.

Poultry

If you live in the Southwest, Bo Pilgrim is probably a familiar figure. Clad in a Pilgrim hat (of course), he delivers TV commercials for Pilgrim's Pride chickens. He also owns 81 percent of the company. His commercials show him as fatherly, gray haired, a bit stern in his Pilgrim hat, and down home, folksy, and full of pride for his product, for his success, and for the quality of the people working for his company. Off camera, he's much the same.

Lonnie A. Pilgrim, who is always called Bo, has been a major participant in an industry that grew from nonexistence to immensity within his lifetime. When Pilgrim's Pride sold 4 million shares to the public in 1986 and was listed on the New York Stock Exchange at $11.50 per share, Bo Pilgrim's 82 percent of the total stock was worth $207 million. Even at that, Pilgrim's Pride was only the sixth-

largest producer of chickens in the country. Pilgrim's Pride sells a little more than half a billion pounds of chicken every year, along with half a billion eggs.

He has been a leader in the poultry industry, and while he has watched and been partly responsible for it, consumption of chicken has gone from 33 pounds per capita in 1965 to 58 pounds in 1985—an increase of 76 percent. In the same period beef consumption went up only 7 percent, pork 14 percent. More importantly, during the same period Choice beef prices per pound went up 191 percent, chicken only 90 percent. The two sets of figures, of course, are not unrelated.

Bo Pilgrim was born in Pine, Texas, just outside Pittsburg, where his headquarters are today. Pine doesn't appear on many maps, and Pittsburg shows a little under 5,000 in population. When Bo was seventeen, he and his brother started a small farm-supply store. They bought feed in bulk and then bagged it to sell to local farmers. Eventually they bought a small feed mill so they could produce the feed themselves instead of buying it already mixed.

The next step up was to go into the newly developed broiler business. They were already selling chicks and feed to customers who paid them if they made money, owed them if they didn't. The Pilgrims decided they might as well take their own risks in growing; they were already taking risks in getting paid. They went on to buy a hatchery to supply chicks, then a plant to process chickens. Bo has been buying or building ever since, making his company a model of the vertical integration that has taken place throughout the poultry industry and is also currently taking place in some other areas of agriculture.

Vertical integration means single ownership, or at least control, of more and more sections of an industry. For chicken producers it has meant control of both quality and costs in a way that was unknown forty years ago, when Bo Pilgrim started. Pilgrim's Pride owns hatcheries, broiler and egg production facilities, feed mills, and processing plants, which not only produce dressed chickens but go on to deboning, forming, battering, breading, and cooking.

Chicks hatching are ungainly sights as they struggle to get out of their eggs. An hour or so later they are confident balls of sturdy fluff.

Broilers,
the Chickens We Eat

Broilers are raised indoors, where everything from feed quantities to ventilation can be controlled. The days of chickens grown "on range" are long gone, victim of high labor costs, lack of quality control, predator damage, disease, and numerous other problems. Today it takes six to seven weeks to grow a mature broiler with a weight of 3.8 pounds, and an average of five flocks per year are cycled through each house.

Broilers aren't just chickens. They are bred to grow fast while converting feed to meat efficiently and with a minimum of fat, as consumers want. Although some growers develop their own broilers, most chickens are bought from genetics companies that specialize in developing and growing broiler chicks.

Housing varies according to local weather. In the North most broiler houses are concrete or frame, totally enclosed to keep out the weather, with ventilation controlled by a complex series of fans. In the South housing is more likely to be open sided, with curtains than can be opened in hot weather, closed to keep out winter drafts. This housing is much lower in cost than the sturdier northern type, and so the concentrated growing areas are in the warm-climate states. Arkansas produces the most birds—in 1985 about 760 million broilers. Next is Georgia, then Alabama and North Carolina. Pilgrim's Pride birds are raised in Arkansas and Texas.

The standard broiler house for Pilgrim's Pride is 300 feet long, the length of a football field, and 40 feet wide, to accommodate 16,000 birds in cool weather, 15,000 when it's warmer. Broilers are grown "on the floor," and the broiler house floors are covered with shavings as bedding or in some areas with peanut hulls.

The ownership and operation of chicken-growing farms varies, but it's rare, almost unknown, these days for individual farmers to finance and raise the broilers themselves, then sell them to processors like Pilgrim's Pride. As the industry has integrated vertically, the processor has become more involved with the grower. Some processors actually have their own "grow-out" farms; other processors contract individual farmers to raise the broilers under very careful supervision.

Pilgrim's Pride does both. It has contracts with some 1,200 independent farmers and also owns about 25 grow-out farms. It supplies chicks, feed, and veterinary and technical services to the contract grow-out farmers, who own their own houses and supply the needed labor for growing the birds. The majority of the Pilgrim contract growers are part-time farmers with a full-time off-farm job. They are paid by the live weight of their birds on an incentive schedule based on a lot of factors, including feed conversion efficiency, death losses, and overall production quality.

Pilgrim's Pride has four feed mills located close to their grow-out farm network and supplies about a million tons of feed a year. The chief ingredients in chicken feed are corn, sorghum, and soybean meal, which have been scientifically formulated by animal nutritionists to give the best-quality bird for the lowest quantity of feed. After mixing at the mill, the feed is trucked to the farms, where it is pumped into storage bins at-

Broilers, chickens raised specifically for meat, are a relatively new breed, separately developed from those chickens bred to produce eggs. Broilers are raised in houses that often are the length of a football field and can hold 15,000 birds.

tached to each broiler house and automatically metered to feeders running throughout the house. The birds have food available constantly.

Broiler producers and processors can change their product much more rapidly than the beef industry—the life cycle of their animals is shorter. When the public became upset over fat and cholesterol, Pilgrim's Pride shifted its feed formulas to grow a leaner bird with less fat, fewer calories, and less cholesterol, and changed its processing to remove additional fat. And Bo Pilgrim's television appearances made the most of it.

Consumer preferences change with breathtaking speed. Of the total chicken sales by Pilgrim's Pride, products that are "value added"—meaning they undergo some sort of additional procedure beyond the preparation of a basic whole dressed chicken ready for sale—went from 56 to 77 percent in three years. The value-added sales include cutting, deep chilling, and packaging in individual trays, as well as deboning, specialty cutting, forming, dicing, battering, breading, and cooking.

These increased services and products are more profitable than the basic bird and help insulate the final price of the product from the vagaries of the cost of producing the chickens—which is mostly determined by the cost of feed. More than that, they are what the consumer seems to want—less time in the kitchen, not only for homemakers, but for restaurants and fast-food chains as well.

Layers, the Chickens That Produce Eggs

Eggs are no longer a growth industry. In 1970 the United States produced 69 billion eggs, before the cholesterol scare stirred the consumer. In 1985 we produced slightly fewer for a larger population. Although not increasing, the number still represents almost 300 eggs per person per year. That means that slightly more than one chicken is laying eggs full-time for each of us.

Layers, bred to produce eggs, normally live out their mature lives in stacked cages, several birds to a cage. Manure drops through the wire floors and is whisked away, and feed and water pass in front of the chickens. As eggs are laid, they roll to a belt that takes them directly to a processing and cooling room. Layers have lent themselves to automation very well.

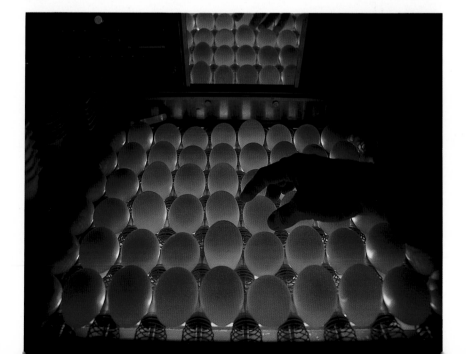

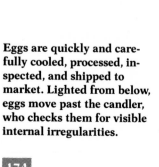

Eggs are quickly and carefully cooled, processed, inspected, and shipped to market. Lighted from below, eggs move past the candler, who checks them for visible internal irregularities.

Although the total production has remained the same, the number of chickens producing these eggs has decreased. As in much of the rest of farming, the production per unit has improved. The average number of eggs produced per chicken in 1970 was 218; in 1985 it was 247. California produces almost twice as many eggs as any other state, with Indiana, Georgia, and Pennsylvania grouped behind.

If broiler production is mechanized, it's nothing compared to egg production. Most of us are familiar with photographs of the batteries of cages holding one or a number of chickens that have nothing to do all day long but eat, drink, defecate, and lay eggs. Food passes in front of them constantly, droppings go through the wire to a collector and disappear, and eggs roll onto a collecting belt. Surroundings are dull, usually dimly lit, but strikingly clean, and eggs hardly settle onto the collecting belt before they are automatically whisked off to a processing room to be washed, cooled, sized and graded, and often packaged for the final consumer.

Layers have been made as much as possible into machines to meet the needs of the producers, and their breeding is aimed very specifically at egg production. They are as different from broilers as Holstein dairy cattle are from Hereford beef cattle. Usually raised in special "pullet" houses, some of which have cages, some of which are open floored like a broiler house, at about eighteen to twenty weeks of age the pullets begin to lay eggs. They are moved to the layer houses and settle in to meet their production schedule. No roosters are needed to induce hens to produce eggs, and therefore market eggs are usually not "fertile" eggs, although there is some demand for fertile eggs in health-food stores.

Pilgrim's Pride's pullets are grown by contract growers, just as their broilers are, but when the birds are old enough to lay, an affiliated company takes the layers into its houses, which have capacity for something over 2 million birds, mostly in houses that hold 100,000 each in three-tiered cages.

Practices vary, but most layers remain in production for about twelve months after being put into the laying house, their egg-laying frequency rising during the production cycle, then dropping off. Toward the end of the cycle, they go into a molt; their egg production drops greatly, they lose some of their feathers, and they become rather morose, if chickens can be called morose. In most layer houses, new birds replace the old, which are sold off for various products. Depending on the price for "spent hens," as they are called, the layers may be kept to lay through another cycle. Their second cycle doesn't match the productivity of the first, but then the grower hasn't had to pay for the pullet stage, when they are growing up and eating but not producing anything.

Much has been made of the confinement of layers—of their boredom, their frustrations, our cruelty toward them. Also to be mentioned is their lack of sex life, for these layers are virgins and have never even seen a rooster. On the other hand, they are fed regularly with the best of feed, have no insecurities, and probably never in their lives have their "fight or flight" mechanisms triggered. The consumer is clearly the beneficiary of all this. The chickens may or may not be, depending on your viewpoint.

Hogs

There's something about a pig. I can't explain it, because pigs are really very ordinary creatures, but show a photograph of a pig and everyone goes wild. For years my photography company has put out an annual promotional calendar featuring pictures of pigs over "cute" captions. For a change of pace, we've tried using other animals—most recently we tried chickens, and I think chickens are incredibly funny—but the calendar evokes no applause. Back to pigs and the response rating soars again.

When is a pig not a hog? The two terms are interchangeable, so far as I'm concerned. If it's necessary to differentiate, a pig is a young hog, but at what stage a pig becomes a hog is mostly in the eye of the beholder. A sow, incidentally, is an adult female; a gilt is a young female before breeding; a bred gilt is a young female that is pregnant. A boar is an adult male; a barrow is a castrated male. Comparing these terms to those used for cattle, you could say that a sow is a cow, a gilt is a heifer, a boar is a bull, and a barrow is a steer.

While beef and chicken have been battling for the American consumer market, pork has quietly held its own. The long-term trend in the amount produced seems about steady or slightly up, although shorter-term numbers rollick along in up and down cycles, the length of which is debated extensively among animal economists. In 1971 there were 95 million hogs slaughtered; by 1975 the number had dropped to 70 million. By 1980 it had gone back up to 97 million, but by 1985 it had dropped again to 85 million.

The number of pigs raised is determined not only by market demand but also by the cost of raising them. A widely used rule of

Funny Pigs

The pig is really not an outstandingly funny creature, but there seems to be something remarkably humorous about photographs of them. Fortunately for the photographer, there is an occasional pig that turns out to be a real ham . . .

thumb is the hog-corn price ratio, that is, the number of bushels of corn required to buy 100 pounds of live hogs. For many years it fluctuated around 20, but in 1987 the price of corn was so low and the price for hogs so high that the ratio reached 40—hog producers were making very good money. As recently as 1984 the ratio was down to 15 to 1. Few prices move as rapidly as farm product prices.

Raising Pigs

Joe Campbell grew up on a farm in central Indiana, studied veterinary medicine at Purdue, and now practices in southern Indiana, where he spends every waking moment in the agricultural world he loves. His wife, Marty, is in charge of his veterinary office and routes Joe and a young veterinarian who has come into the practice on their daily visits to local farms. About 80 percent of their veterinary practice is in food animals—beef, hogs, sheep, and chickens. The rest is small animals and horses.

The practice includes a number of swine operations. Many of the hog raisers in the area also have beef herds. As Joe explained to me, "A lot of this has to do with manure. Hog manure here is mostly spread on pasture grass, although some is knifed in. Given the very positive effect of the manure on pasture, there has to be some animal to use the increased pasture grass. Beef cattle are a natural." Joe and Marty raise a purebred herd of Saler cattle of their own.

Most of the farmers in the area are part-timers. They work in town at full-time jobs and still manage to be farmers, some of them on a surprisingly large scale. Cow-calf herds, which mostly graze pasture, require rela-

tively little management time or labor. While hog operations are more labor intensive, mechanization reduces the labor.

Joe's philosophy is to practice preventive medicine rather than to run a "fire engine" practice. "If we can prevent problems from occurring," he says, "rather than wait for them to occur and then try to cure them, we cost the client less. We get him a more healthy product for less money.

"It's pretty clear from studies, as well as from my own experience," he continues, "that the larger producer actually invests less money per animal in veterinary services. The larger producer is the one more interested in prevention."

The hog operations in Joe's area are mostly "farrow to finish"; that is, the sows farrow their pigs, and the young pigs are then raised to market weight on the same farm. In some other parts of the country, operators of farrowing setups sell off their young pigs at about eight weeks of age, weighing 35 to 40 pounds, to specialists who feed them to market weight but do not do their own breeding.

The days of sows wandering around an oak forest, digging in the mud, are pretty well gone. Most hogs these days are raised in confinement, totally inside a building, or in semi-confinement, which usually means they are in a building with one end closed, the other open. Confinement buildings may have concrete floors, but many have floors made of steel mesh, often plastic coated, which allow droppings to fall through and be automatically washed away.

In confinement growing operations, boars are kept in segregated quarters except when they are moved into a special area for breed-

Today, most sows are placed in a farrowing crate about a week before they give birth. They can stand up or lie down but not turn around. After the baby pigs are farrowed, about eleven to a litter, they can escape from under the farrowing crate bars to avoid being rolled on, but they can still reach the sow to nurse. The sow and her young pigs remain in the pen with the farrowing crate for three to five weeks, when the pigs are weaned and moved into nursery units.

ing. When pregnant, sows live in a gestation area that may have individual pens or may have "open housing" holding a number of sows. It takes a little over a hundred days from breeding to birth, and hog producers try to get at least two litters of ten or eleven pigs from the sows each year. A sow that produces twenty-five pigs a year is doing very well.

About a week prior to their farrowing time, the sows are moved into individual stalls, or farrowing crates. These vary greatly in design, but the general purpose is to keep each sow and her litter separate from the other sows and litters. Not only are the stalls designed to allow the young pigs to stay away from other litters, but they also protect them from their own sow's activities. A sow is a lot of animal, and when she lies down, her pigs need to be out of the way.

Young pigs, which weigh 3 to 4 pounds at birth, are astonishingly capable. Born with a sheath, or amniotic sac, around them, they fight their way out of it and immediately go around their mother, search out a teat, hook on, and begin to nurse. It is an astonishing sight to watch. The young pigs stay on their mother for three to five weeks. The age of weaning is a big point of difference among swine experts, but after weaning the pigs are removed from the crate and put into a nursery unit. Pigs that come off the sow at three weeks weigh maybe 12 pounds or a little less; if they stay on the sow for five weeks, they may weigh 15 to 20 pounds. They grow pretty fast.

Usually the young pigs are sorted by size in the nursery, with pigs of similar weight put in group pens, and they may stay in the nur-

sery for three to six weeks. Then they likely are moved again, into a grower type of building with bigger pens and more pigs per pen. Finally, they go into group pens in finishing buildings where they are fed until they are ready for market, when they weigh 220 to 240 pounds. It takes them five to seven months from farrowing to reach that weight.

A hog's diet may vary considerably, depending on the cost of various feeds, but corn is the most basic feed. Sorghum is sometimes substituted for corn, and other grains such as barley and wheat may also be used if the price is right. Soybeans in various forms are used as a high-protein supplement. During the stress period when young pigs have just been weaned, many producers give them some form of milk, either dried whey or dried skim milk. Animal nutrition these days is very carefully controlled, completely recorded, and closely studied.

The sow, after her pigs have been weaned, goes back to a breeder unit and is rebred to begin the cycle again. How soon she is bred depends on the individual grower's ideas, but it's likely to be pretty quickly. Time in a sow's life is money for the farmer.

The veterinarian tries to have input throughout the whole process of growing. Joe Campbell explains, "The age most susceptible to disease and death is probably the farrowing period, but there's a stress period when the pigs are weaned. We try to get the pigs started well on their own, happy and growing. If we can get them off to a strong start here, they have a good chance to do well for the rest of their lives."

As in most of the rest of agriculture, there's considerable speculation about whether the independent hog farmer is going to be taken over by giant corporations and disappear from the scene altogether. Joe Campbell feels it's unlikely, although he is certain the number of hogs per farm will increase and the number of farms devoted to hogs will continue to decrease. From 1978 to 1982 the number of farms with hogs dropped by more than one-fourth, but hog numbers dropped by only 4 percent.

There are a number of big operators in hog

These pigs have been weaned at four weeks and have been moved to group pens in the nursery, where they will stay three to six weeks. Like calves, which are usually sold to be finished in a feedlot, many pigs that have been weaned are sold, often through auctions, to be raised on a farm that doesn't breed pigs but only finishes them. Some farms, however, are set up "farrow to finish"—they breed sows but also raise the pigs to market weight.

farming, with thousands of sows producing young pigs year round, and these big operations usually plan to get even bigger. Some of the big operators own all their own hogs, from farrow to finish, while others contract with independent farmers to raise the hogs, much in the same way broiler chickens are generally grown. The trade continues to debate the pros and cons of being small, of being big and owning the animals, of being big and contracting out the animals. My feeling is that for the foreseeable future there will be an evolving mixture of production techniques, but it is too early to determine the direction this evolution will take. When pork prices are high, bigness seems wonderful; when prices sag, as they always do, bigness can run up huge losses quickly.

New trends in nutrition, new housing concepts, new breeding goals, and even owner-ship shifts in the works. One major innovation that could have a profound effect on the industry is porcine somatotropin, the synthesized version of a natural enzyme occurring in hogs. Its use can increase the rate of weight gain on fattening pigs by about 15 percent. In addition, the meat resulting from the use of somatotropin carries far less fat—a goal hog breeders have been working toward for twenty years. At this stage of development there seem to be few side effects on hogs and none on humans who consume the pork. Although porcine somatropin has not been approved for use by the government, it is close enough that pharmaceutical companies are building plants to manufacture it. A major breakthrough like this is certain to upset the hog feeder's routine and perturb the markets, but ultimately it should benefit the consumer.

Dairy Cattle

Only a few decades ago, Mr. Stewart's dairy was like many others across America. His glass-bottled milk would appear magically every morning on our doorstep in the small Pennsylvania town where I grew up, and often in the late afternoon I would ride my bike out to the dairy and watch the milking. Mr. Stewart knew each cow by name, of course, and would slap each one on the rump as he sat down on the milking stool beside her, talking quietly, maybe to her or maybe to himself—I never knew. If he occasionally spat on his hands before he started milking, I never noticed. I marveled as he sometimes deflected a teat and shot a squirt of warm milk directly into the mouth of one of the cats that always sat nearby watching and hoping.

There are not many dairy farmers left today who routinely milk cows by hand, as Mr. Stewart did. The average dairy in 1940 had 5 cows; by 1982 the average had soared to 39, but California's average was already more than 200 cows. Dairies in California that milk 2,500 or 3,000 cows are not unusual.

Dairying has followed the standard agricultural pattern. There are fewer dairy farms each year, and each farm milks more cows using more and more automation. The average production of milk per cow in the United States for 1971 was 10,015 pounds, which by 1985 had increased to more than 13,000 pounds. California managed to boost its average production to almost 16,000 pounds. That's a lot of milk, but the record holder is a cow that produced more than 55,000 pounds of milk!

A cow first produces milk after she has had a calf. A female calf, called a heifer, can be bred when she is about fifteen months of age. She will weigh about 800 pounds by then,

Although dairy calves can nurse, they usually are separated after birth from their cows. They are fed the milk from their mothers, since it contains special substances they need, but they get it from a bucket with a rubber teat.

The amoun
cows produ
with the na
more than
year. Califo
the most p
more than

185

having g
After a g
for a hu
starts to
present
the cow
fed milk
early m
young ca

After h
a cow, a
mature,
if she i
breed. A
conceive
that has
a few we
ues to g
her nex
longer r
new cal
born cal

Cows
the norr
six year
cow is r
in the
younger
product
aren't p
ber of
mained
years, b
gone up
ter man
agemen
knowle
With th
produc

186

Breeding Dairy Cattle

Unlike breeds of dairy cattle, breeds of beef cattle in the United States have increased almost exponentially; there are dozens of recognized breeds of beef cattle, with new ones coming into and going out of popularity with increasing regularity. In dairy cattle the opposite has happened.

The Holstein breed has simply overpowered the other breeds. It accounts for more than three-fourths—some say as much as 90 percent—of the total dairy cattle. There are several reasons for this. First of all, the Holstein was already an excellent cow when the first ones were imported in the mid–1800s from Holland. Second, its breed association has done a superb job of improving a good beginning. Also, when consumers in the 1940s and 1950s began switching from butter to oleomargarine, first because of price and later because of the health concerns about the consumption of fats, the Holstein proved to be the right breed. Holstein milk is lowest in butterfat, with an average of 3.7 percent; Brown Swiss average 4.1 percent, Guernseys give 4.8 percent, and Jerseys, the diminutive cattle with the doe eyes, top the list at 5.2 percent. The Holsteins have less butterfat, but they give considerably more milk per cow than any other breed.

One of the major contributions to increased milk production has been artificial insemination, usually called AI. Like some science fiction phenomenon, AI actually allows the breeder to know the proven quality of a bull through studying its offspring *before* the bull is generally used. It is like being able to see a child at age twenty before he or she has been conceived.

It actually takes about five years to find out if a bull is going to be a desirable sire. Potential sires are first old enough to be bred when they are twelve to fourteen months of age. Breeding groups (both cooperatives and private organizations) collect enough semen—they use an oversized condom to collect it—from the new bull to breed about 600 cows. About half of the resultant calves are, of course, male and are not used in further testing. But the females, or heifers, are bred at about fifteen months of age, and after calving they are into their first lactation, or milking period.

Great quantities of figures on the milk production and other traits of the offspring of the test bull are accumulated. The Department of Agriculture does the actual collating of the information, which comes to them from a variety of sources, and twice a year issues "report cards" on bulls whose progeny are under test. The breeders then sit down with this and other information and decide which bulls should be added to their string. The lengthy test procedure results in less than 20 percent of the bulls tested becoming sires.

The bulls that are put into service have their semen collected regularly, diluted, and frozen in "straws" immersed in liquid nitrogen. Because the semen can be kept viable almost indefinitely, a single bull can inseminate as many as 10,000 cows annually, instead of, through regular breeding, fewer than a hundred. And the dairy farmer, by a simple phone call to his breeding service, has a choice of semen, at very low prices, from dozens of bulls.

There are other interesting developments in the offing. Already in use, with about 100,000 calves a year being born from it, is

The science of embryo transfer has markedly improved the quality of dairy animals. It allows superb cows to produce more calves with their genetic traits than would be possible with ordinary breeding.

The calf on the left was actually born from the cow shown. The other calves are the result of eggs produced by the cow, flushed from her, and transferred to other cows. The same bull was used to fertilize all the eggs, so the genetic inheritance of all these calves is the same, even though they were born to different cows!

embryo transfer, abbreviated as ET. It's expensive enough that it can't be done for every cow on every farm, but it's been proven worthwhile for the highest-quality animals. Essentially, ET permits a number of eggs to be removed from one very high-quality cow and implanted into lesser-quality cows for development into calves. The genetic makeup of the resulting calves comes from the original cow and bull.

Why go through this complicated procedure, wonderful as it sounds? Because even though the best cow normally produces only one calf a year, through ET she can produce a great many more (figures on just how many are changing rapidly), and genetically better cows produce better offspring that are capable of producing more milk. And, unlike somewhat similar procedures in humans, cows don't get involved in lawsuits over parenthood.

Another scientific breakthrough doesn't involve breeding techniques and isn't yet in use, but it is far enough along that factories are already being constructed to manufacture it. The product is bovine somatotropin (shortened, of course, to BST), a synthetic version of a naturally occurring protein produced in the pituitary gland of cattle.

BST simply increases milk production by between 15 and 20 percent. What it seems to mean is that if a farm wants to keep its production the same, it can use BST and do with 15 to 20 percent fewer cows. Or it can produce 15 to 20 percent more milk with the original number of cows. Although BST has not yet been approved by the government for general use, there seem to be no hitches to its acceptance. Except, of course, milk is already being overproduced and this will add one more complication to the efforts to control production. So it goes in the scientific world.

One of the solutions to the problem of overproduction of milk has been, from 1985, a government program called, significantly enough, the Dairy Termination Program. Because there were too many dairy cows, dairy farmers were given the opportunity to sell some or all of their cattle—either to become beef in the supermarket or to be exported. Paying the bill was done partly by the dairy industry itself and partly by you and me.

There were really big bucks involved. In California, more than seventy-three dairies sold enough cattle to get paid more than $1 million dollars each; twenty-one dairy operations in Florida got over $1 million. There were 1.1 million head of dairy cattle slaughtered, and another 60,000 shipped abroad under the program. The total cost isn't yet in, but in spite of a 1.7 percent drop in cow numbers in the first year, total milk production still went up close to 1 percent. Cows that were not "terminated" obviously worked harder.

One group of losers from the Dairy Termination Program was the beef cattlemen, since the slaughtered dairy cows ended up going through beef channels and pushing beef prices down. The government, alarmed by shouts of anguish from the beef industry, bought up 400 million pounds of beef to offset the Dairy Termination Program and about half of this was sold very cheaply to Brazil.

Sheep

Generalizations in the sheep business are dangerous. The hobbyist in the East who has a handful of sheep that he shows at the local county fair and pastures on thick grass has a different approach from Buster and Linda Duferrena, who run 1,800 ewes in the tough, dry country of northern Nevada. For the Duferrenas their sheep operation means making it or going broke, although Buster does hedge by also running a 400-cow beef herd. It's a beautiful way of life, and three of the Duferrena children are working on the ranch, but it takes hard work to make it in the sheep business.

Of all the seasons, the Duferrenas prefer lambing time in early April, when the sheep are close by the ranch, trailed in on foot from winter pasture down on the Black Rock Desert, a hundred miles to the south. While the ewes wait to begin lambing, they are sheared

of their wool. Most American sheep are raised both for their meat and their wool. Even with income from both wool and meat, the sheep business has for a long time not been a very profitable one.

Shearing at the Duferrena ranch is done by a professional crew of about ten people who come in to shear the 1,800 sheep and then move on to another ranch. It's difficult, tiring work. The ewes have to be handled as gently as possible, and the shearer must have strength and agility to hold a struggling ewe, one arm and both legs controlling her, while running the electric clippers. As the clippers slide along the skin, the wool comes loose in large patches. The Duferrena ewes, mainly Rambouillet and Colombia crossbreeds, produce about 10 pounds of wool per sheep, somewhat higher than the average.

Lambing is scheduled for the fifth to the

twentieth of April. While lambs weigh about 10 pounds at birth, the development of the lamb embryo is unusual. The gestation period of ewes is five months, but at four months the fetus weighs only a pound; the other 9 pounds are developed during the last month of pregnancy.

Lambing is interfered with as little as possible. The ewes are let loose "in the brush," as Buster Duferrena describes it, about 5 miles from the ranch, in a sheltered area that has been selected because it is out of the north wind and has space for each sheep to lamb separated from the others. At lambing time weather can still be a problem. Buster shrugs his shoulders and says, "Well, you hope for the best."

Under these range conditions the Duferrenas count on having more than a 100 percent crop—that is, more live lambs than there are ewes. While a flock this large always has barren ewes that have no lambs, twin lambs are not uncommon. Smaller flocks in richer eastern country are handled so that ewes can average almost two lambs per year.

The Duferrena ranch is a long way from anywhere. When I asked Linda how to find their place, she assured me it was very easy. "Just turn north out of Winnemucca," she said. "Drive 80 miles and look for a mailbox on the right, then turn left down the gravel road about two miles and you'll find us." I did. This is spectacular dry canyon country with brush but few trees; the tops of the mountains to the west are green with forest. It's what most of us think of when we think of western sheep country.

When the lambs have about a month of growth on them, around the tenth of May, the ewes and lambs are "trailed" on foot the 20 miles to their summer range in the nearby mountains. The trip takes two or three days, with the herders and their sheepdogs moving the ewes and lambs slowly. The older ewes know the trail, and the movement seems to be in slow motion unless there is some emergency. It is the job of the herders and the dogs to see that emergencies don't arise. The ranch is 4,200 feet above sea level, and the sheep may even go as high as 7,000 feet for their summer pasture.

The sheep need constant managing and protection and the herder must move them to new ground every couple of days because sheep are notoriously hard on grazing land. Their tiny feet dig into the ground, their sharp teeth clip the grass short. The herders must also be constantly on the lookout for predators, which in the mountains are most likely coyotes and bobcats. Although the area has a high population of eagles, Buster says they seem not to be much of a problem. It is estimated that predators cause close to $100 million in sheep losses each year, giving rise to a bumper sticker that says, "Eat Lamb—10,000 Coyotes Can't Be Wrong!"

Buster Duferrena's father came to the United States from Spain—they are Basque in origin—to herd sheep and stayed on to become a citizen. Like Buster, many of the sheep ranchers in the West are first- or second-generation immigrants, mostly Basque or Greek, who learned the business, became citizens, and bought their own herds. For many years, both Basques and Greeks were able to move to the United States under special visas because they had experience as sheepherders, and in America sheepherders are hard to find.

The job of the sheepherder is surely one of

Most of the large commercial sheep herds are in the West, where they can take advantage of poor—but inexpensive—pasture. Lambing usually occurs in the spring, and the ewes are sheared of their wool around lambing time. The ewes and lambs go to mountain pasture for the summer, then are moved back to the ranch for fall breeding. They may be kept around the ranch during the winter or taken somewhere else for winter pasture.

the loneliest still remaining, even if the old horse-drawn sheepwagons have mostly been replaced by house trailers pulled by pickups. It's a job that few present-day Americans want or can cope with. A rancher once told me of his experiments with hiring school-teachers on summer vacation to run his sheep in the San Juan Mountains of Colorado. "They all thought it would be great to be alone in the mountains after a winter of noisy kids," he said, "but none lasted more than a month, and most came trailing back down to the ranch after a week. I guess they just couldn't stand themselves that long. I import Mexican herders now."

Buster gets his herders from Spain, Mexico, Peru, and Chile. Under a special contract our government permits them to stay three years;

then they are required to go back home but may return for additional stays. There are about 800 herders in the United States under this program. Without them it is very doubtful that the western sheep business could ever survive.

After a summer on pasture, the lambs are separated from the ewes and brought back to the ranch in September weighing something over 80 pounds. The ewes are kept on the inexpensive mountain pasture until the last week of October. When they are brought back to the ranch, the rams are put in with them for breeding. One ram services about fifty ewes during a forty-five-day breeding period.

When the lambs are brought down from summer pasture, the heavier lambs go directly to market, and the lighter ones are

Sheep on pasture must be constantly tended, both to protect them from predators and to keep them moving, for they are destructive of the grass. Many of the sheepherders today are from Mexico or South American countries and work here on special visas. Americans apparently find the job too lonely. After weaning in the fall, lambs may go to feedlots, much the way that beef calves are put into feedlots.

often trucked to southern California, where they are fed on alfalfa pasture to market weight, or to Colorado, where they are fed in feedlots until ready for market. In the feedlot—much like a cattle feedlot but of course on a smaller scale—lambs can gain a little better than half a pound a day, and they go to market at about 120 to 135 pounds.

After breeding, the ewes are trailed about a hundred miles south to the desert for the winter. In addition to government ground he leases there, Buster leases 14,000 acres from the Southern Pacific Railroad. Pasture is sparse—there scarcely seems to be anything for the ewes to eat—but the acreage is huge, and sheep are the best scroungers of our domestic animals. Again, the herders are constantly with the sheep. Thanks to their heavy coats, sheep do well in cold-weather conditions if the snow is not too deep. Come spring, they are trailed back to the ranch in time for shearing and lambing, ready to begin the cycle over again.

As is true in most other areas of agriculture, the government is deeply involved with sheep raising. There might not be a sheep industry if that weren't the case, for the number of sheep in the United States has been dropping steadily for the past fifty years. The total number of sheep in 1971 was about 20 million; by 1985 there were fewer than half that number. There is considerable belief the number is again increasing, but to date I've seen no statistics to bear that out, although the spring of 1987 did see the number of replacement lambs increase by 23 percent over 1986. Whether the market will be able to cope with the lambs resulting from that increase has yet to be determined.

Without government protection and government payments, the situation would be far worse. The protection comes in the form of a levy that has been imposed on lamb imported from New Zealand in response to our conclusion that New Zealand was subsidizing its lamb to sell it at a lower price. The tariff hasn't stopped the imports, but it has made the imported meat higher priced and more competitive with our own. How long the levy will be in effect is impossible to guess.

Incentive payments from the government are really what keeps the American wool business alive. In 1985 the average price paid on the market for shorn wool was only 63.3 cents per pound, but incentive payments from the government came to $1.02 per pound, so growers got 60 percent more from the government than they did from the market! Taxpayers will be happy to know the incentive payment money comes from a tariff imposed on wool imported into the United States. Incentive payments to American producers are limited to 70 percent of the total tariff, although they usually come to less than half of it.

Sheep flocks in the United States can't be considered a major industry. While we consume almost 80 pounds of beef per capita annually, and about 60 pounds of chicken, we eat only a pound and a half of lamb and mutton. And we imported more than 34 million pounds of lamb and mutton from Australia and New Zealand where sheep raising is a major industry. We have only 6 percent as many sheep as Australia. Buster Duferrena is counting on that to change, but whether he will be right remains to be seen.

Vegetables, Fruits, and Nuts

They are thought of together, probably because that's the way we see them in supermarkets—oranges next to cabbages next to spinach next to grapes, and almonds down the row. Even if we want to separate fruits from vegetables in thinking about them, it is surprisingly difficult. One definition widely used is that a fruit is consumed with dessert, vegetables with the main part of the meal. My practical definition is that vegetables are the edible parts of plants that are annuals or at most biannuals. Fruits are the edible parts of plants that are kept in production more than two years. I've never seen that definition in a book, but it will suffice.

Fruits are likely to be grown on trees in orchards or groves, although grapes are grown on vines in vineyards, and so are kiwi, also known as kiwifruit. In the United States, we say that kiwifruit is grown in vineyards, but in New Zealand the plantings are called orchards. Nuts are usually grown on trees, in groves. Some nuts such as pecans and black walnuts are gathered in part from native trees, but most nut crops come from trees cultivated in groves. Peanuts are not really nuts, although they look like nuts, are called nuts, and taste like nuts. They don't, however, grow like nuts; in fact, they grow under the ground like potatoes.

Frequently, fruits and vegetables are marketed similarly, and many have in common being marketed both fresh and processed. Some, like lettuce and watercress, come to us only fresh; olives and raisins come only processed. Most fruits and vegetables, however, are available fresh at certain times of the year and as frozen, canned, or dried products throughout the year. Nuts, which have much longer keeping qualities than fresh vegetables or fruits, may be available fresh but often have some processing. They are also sold in cans or glass.

We consume great quantities of all of them. If we calculate by their weight as they leave the farm, we each eat an average of almost 200 pounds yearly of fresh and processed vegetables, a little less than half of them fresh, almost the same percentage canned, and 10 percent of them frozen. These vegetable figures do not include potatoes because potatoes are, by themselves, such a large figure: 125 pounds per person. We also eat about 125 pounds of fresh and processed fruit, more than 70 percent of which is consumed fresh;

In the good old days there were roadside vegetable stands everywhere, as there still are in Lancaster County, Pennsylvania. There are some indications that roadside stands are making a comeback, but their share of the total vegetable or fruit market is too small to be counted.

by contrast, we each eat less than 3 pounds of nuts (not including peanuts, which aren't nuts; we eat about 7 pounds of those).

We eat different parts of the plant in different vegetables. Think of the vegetables whose leaves we eat, such as lettuce and spinach, but we eat the bulbs of onions; roots of beets and carrots; the tuber of potatoes; the growth stem of asparagus; the flowers of cauliflower, broccoli, and artichokes; the seeds of peas and beans; the fruit of tomatoes or peppers; and the immature fruit of eggplants and sweet corn. We probably don't think of that

when we see them in the market; they are just vegetables.

In fruits we eat the fleshy area surrounding the seed or seeds and sometimes the seeds as well. If that is the case, then why are tomatoes called vegetables? After all, we eat the fleshy area around the seeds of tomatoes. It seems to me it is because they are annuals, or at least they are annuals as we grow them, although they can be perennial plants. Artichokes also cause problems for my definition. They are perennials, but I've never heard them called fruits.

Vegetable, fruit, and nut harvesting is still in the beginning stage of mechanizing. Facing page: Pecans (left) can be shaken from the trees and collected mechanically. Cantaloupe pickers (right) have to decide which fruit is ripe, then pick it, lift it from the ground, and carry a heavy bag of cantaloupes to a truck. A field may be gone over several times.

It takes a strong arm and back to cut cabbages from the stalk, trim them, and put them onto a collector belt. The belt moves the cabbages to packaging equipment mounted on a truck that moves through the field with the pickers.

The marketing of fresh produce is done in many varied ways. There are large marketing companies involved in supplying fresh produce throughout the country, often on a year-round basis. However, because of the premium placed on freshness, there are many regional producers and marketers.

One of the interesting revivals has been a resurgence of activity in "farmers' markets." These are usually roadside markets, right at the farm, a pleasant drive from the city, or they are a space in town, often a school parking lot so the market can be held on Saturdays. The Pennsylvania Dutch have a long tradition of "tending market," and their markets have now spread from the cities in their own countryside to Philadelphia and even to New York and Baltimore, and their products have expanded from homegrown produce to home-baked bread and homemade crafts. The cities extend a warm welcome.

Another area of vegetable and fruit growing is home gardening. Even in the heart of the cities there are hundreds of enthusiastic gardeners. New York City, for example, has a major program of "urban horticulture," and its Cooperative Extension's monthly newsletter goes to 5,000 readers. There are no statistics on farmers' markets or home production, and they probably represent an insignificant part of the overall marketing picture for fruits or vegetables, but to me they are a welcome relief from the boredom of bigness.

Farmers' markets can exist only because of the large markup marketers must make to pay for the transportation of many fresh fruits and vegetables from long distances. While many cities are surrounded by areas that can grow produce for local consumption during a relatively short growing season, the bulk of fruits and vegetables for most of the year must be delivered over long distances at high speeds. An intricate network of salespeople and brokers manages the process, connecting producers with buyers, but it is our trucking system that gets the fresh produce to market in good condition. The figures show that transportation costs are sometimes greater than the cost of the products themselves. The demand for freshness is more than anything else a mark of our affluent society; we are willing to pay for it.

Vegetables and fruits grown for processing are frequently grown and handled differently from produce for the fresh market. Tomatoes are possibly the best example of this. Fresh tomatoes must be handpicked. They need to be large and without blemishes, and they must be picked before they are fully ripe so they can withstand shipping and be ripe on the market shelf. A field may be gone over several times because the tomatoes ripen at different times. Canning tomatoes, on the other hand, have been specially developed so they ripen simultaneously. They are purposely smaller and more solid, to withstand mechanical harvesting. When the tomatoes are beautifully ripe, the entire vine is lifted from the ground by a harvesting machine in a single operation, and the tomatoes are separated from the vines and sometimes even selected for ripeness by color. They are loaded into trailers and arrive at the processing plants in less than an hour. The harvest of processing tomatoes is a remarkable race for preservation of quality.

The growers of vegetables have a different philosophical makeup from fruit or nut pro-

Fruit growers, unlike most vegetable growers, are looking at a lifetime investment in their trees. It may have taken five to ten years for this plum orchard to produce fully, whereas a vegetable grower can have a crop to sell in a few weeks. Capital becomes vitally important in the fruit business.

ducers. Vegetable growers can generally look at immediate results and almost immediate payment for crops that may take as few as sixty days to produce. Vegetable growers may even be itinerant producers who use leased land and follow the sun, producing crops wherever there is a season to do it. Fruit and nut growers are just the opposite. They need long-range planning and massive capital. They usually don't recoup any of their investment for at least three years, and once committed to fruit, their orchards or groves are a lifetime project. Frequently orchards are torn out not because they have quit producing but because new varieties or new fruits have become more popular. Fruit growers need long-range planning and massive capital.

Beating the market—getting to market earlier than your competitors—is a way to get premium prices for both fruits and vegetables. With vegetables, one technique is to start a crop under hot caps, which allow the sun to heat the soil and start germination but protect the young plants from cold spells.

Vegetable Crops

The vegetable crops in the United States reflect how bountiful and varied this country really is. They also reflect our national origins, for we have all kinds of vegetables that once were of interest only to ethnic markets but now are generally available to a wide market—for example, chile peppers, artichokes, and greens.

Vegetables vary a great deal in how they are grown and how they are marketed, and so there's no general pattern. The one factor that really stands out is that California dominates vegetable production, producing about half of the total fresh and processed vegetables grown in the entire country.

Take a tour through the Salinas area south of San Francisco and you'll find all kinds of vegetables in all stages of growth. The narrow back roads around Salinas are clogged with trucks and trailers moving from the fields or to the fields. Reefers—refrigerated trucks—fill the main highways as they rush perishables to market. The feeling is one of frenzy.

The principal vegetable crops grown in California occupy twice the acreage of the next-largest producing state—which, surprisingly, is Wisconsin, with Minnesota third and Florida fourth. Wisconsin's acreage is almost entirely devoted to vegetables for processing, however. Florida, known for winter vegetables that are distributed to the urban centers of the East Coast, has almost one-fourth as much fresh vegetable acreage as California. Florida is just the opposite of Wisconsin, since almost none of its vegetables go for processing.

Excluding potatoes, a little less than half of the total vegetables grown are marketed fresh, and slightly more than half are pro-

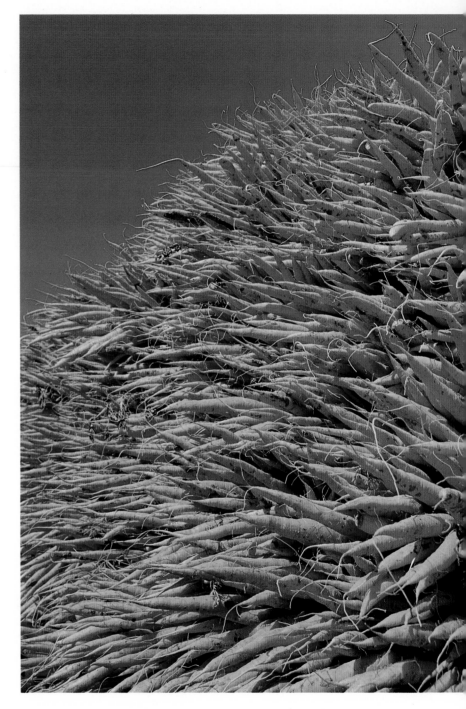

Massive as they seem in the field, pumpkins are really only a minor vegetable. Still, 20 million pounds of pumpkins and cooked squash are commercially frozen each year. It is nearly impossible to realize the incredible amounts of vegetables we use. Almost 100,000 acres are planted to carrots, and we consume more than 2 billion pounds of them a year, about two-thirds fresh, one-third processed. We eat twelve times as many pounds of potatoes (right) as carrots. Like carrots, the edible part of potatoes is grown underground.

cessed. About 10 percent of the processed vegetables are frozen, and more than 40 percent canned. Figures on potatoes are usually separated from those on other vegetables because they skew the numbers. Sweet corn is the second most popular frozen vegetable, but we eat only 15 percent as much frozen corn as frozen potatoes.

Looking at three vegetables—lettuce, potatoes, and artichokes—you will see how varied vegetable farming can be. Lettuce is entirely a fresh vegetable, requires hand labor, is grown in many places, and must be rushed to market. Potatoes are a mechanized crop also grown in many different areas, have excellent keeping qualities, and are consumed in unbelievable quantities, both fresh and processed. Artichokes are grown in a tiny area of California and are an example of a specialty crop about which most of us know very little.

Potatoes

The best thing about potatoes is that they can be processed in so many ways. Each of us consumes almost 125 pounds of them each year. If you're thinking just of baked or boiled potatoes at home, you may be suprised to hear that these account for less than half, about 41 percent, of the total pounds. Frozen products on the supermarket shelves account for another 34 percent, and 8 percent is dehydrated. Then there are chips and the potatoes needed for canned products such as soups or stews. It adds up to a lot of potatoes.

Raymond Erikson is a farmer on the Rexburg Bench, a flat potato-growing area above the town of Rexburg, Idaho. His grandfather, Alex, came to Rexburg about 1896, and the

family has been farming and expanding their acreage ever since. Raymond and his brother farm together and maintain they aren't big growers, but they farm about 250 acres of potatoes. That means in an average year they harvest about 7 million pounds of potatoes.

Potatoes grow well in cool climates, and all of the states that lead in potato growing are on the Canadian border. Idaho leads in potato production, Washington is next, and Maine is third, although Maine has only one-fourth of Idaho's production. The potatoes used for planting the Rexburg Bench crops are usually grown higher in the mountains, where the weather is cooler and there are

Potatoes (top right) are planted using cut-up potatoes as the planting stock. Potatoes grown from seed have greatly varying characteristics, but those grown from vegetative parts will all be quite similar. Certain varieties of potatoes have a lovely bloom (left); others have tiny flowers. A potato harvester (bottom right) picks up potatoes that have already been dug and are lying on top of the ground in rows. It removes the dirt from them and transfers them to trucks that move alongside the harvester as it goes through the field.

Sprouting Seeds

One of the agricultural miracles, the most basic one, is the fact that seeds sprout and grow. They do so in varied ways.

Facing page: Corn plants (left) start from seed but potatoes (right) usually are planted from a section of a potato.

A bean germinates (right).

Lettuce

Regardless of the time of year, you can always find iceberg lettuce in the supermarket. Iceberg lettuce is grown in an astonishing number of different parts of the country at different times of the year, and so it is always available and its production is far higher than that of any other type of lettuce. Iceberg keeps well, grows in compact heads so it requires little trimming or other field work, and has been developed in different varieties for differing climates and soils.

Many perishable vegetables, including lettuce, stay fresh much longer if they are cooled rapidly after they are picked. In vacuum cooling, a truckload of lettuce that has been packed into cardboard shipping cartons in the field, maybe 300 cartons of 24 heads, is loaded into a steel chamber where the atmosphere is pumped out until the air pressure is about the same as it would be in a plane flying at 80,000 feet. The vacuum evaporates excess water from the lettuce, and the lettuce, still in the chamber, is quickly refrigerated. To cool lettuce coming from the field at 75 degrees to just above freezing, the vacuum cooling cycle takes only about fifteen minutes. The lettuce is then loaded into refrigerated trucks, which vastly improves the lasting quality of the lettuce and makes long-distance shipping possible.

Vic Smith is in his thirties, has a degree in business instead of agriculture, and operates a complicated family-owned agribusiness with his father. They are not primarily growers, but are vacuum coolers, and lettuce is the main crop they cool. Their company operates one cooling plant in Colorado, three in southern New Mexico, six in Arizona, and they also operate in the big lettuce areas of California. The growers are their customers, and since the cooler operator gets paid a fee for every carton cooled, this part of the Smith business is not as risky as actually growing the crop. Almost all of their employees are hired on a year-round basis and move from one plant to another as the lettuce grown in different areas matures.

This explains why there is fresh lettuce in the supermarket all year long. The lettuce you buy in late summer might come from the San Luis Valley of Colorado, for example. Fall lettuce might come from Las Cruces, New Mexico, or in the late fall season from the San Joaquin Valley of California. California's Imperial Valley and southern Arizona produce during the winter, and both the San Joaquin in California and southern New Mexico have a spring season as well as one in the fall.

On the East Coast, Florida has a winter production season, and its product competes in the eastern cities with lettuce coming from Arizona and California, although Florida has the advantage of shorter transportation to the northern urban markets. States that have immense population, like New York and Illinois, also have a short summer growing season for local consumption. Transportation costs from distant places become a major part of the consumer price. A carton shipped from California to New York might cost more for transportation than for the lettuce itself. Still, New York State's production comes to less than 2 percent of California's.

In addition to having a cooling plant in the San Luis Valley of Colorado, the Smiths' company actually grows lettuce there as

Lettuce is a tricky crop to grow, particularly because of its water needs, but it does well even in Alaska.

well. They also grow cauliflower, broccoli, onions, and a few other vegetables in Mexico. (Crops raised in Mexico are likely to be those that involve hand labor, since labor is much less expensive there.)

Lettuce is a touchy crop to grow and is very sensitive to the amount of sunlight available. During the summer in the San Luis Valley, when the days are long and hot, lettuce produces a crop in sixty-five days, but during the winter season in Yuma, when days are short and the weather is cooler, a crop may take twice as long to mature.

Lettuce is grown as a row crop and carefully tilled and sprayed with herbicide to minimize weeds. Because it grows so fast, it must be very carefully tended and kept under control. Like every crop, lettuce has its touchy times. As Vic Smith explains: "About the time the lettuce is still very small, with only three tiny leaves out of the ground, it looks as if you've got to keep water and fertilizer going or the lettuce will never make it. Instead, you have to do the opposite. You stress it by withholding water to make the roots go down into the ground searching for water. If you come in too fast with water, it'll never develop the proper root structure, but if you don't get enough water on, just right, you'll kill the plant." He pauses to make his point: "You can't watch that through some car windshield driving down the road; you have to walk the fields to see precisely what's happening. That's the kind of crop lettuce is."

Harvesting lettuce is mostly hard work, sometimes helped by machinery that pickers follow through the field, putting the lettuce into boxes on the moving machine. But this machinery only helps with the loading; it is still necessary to stoop over in order to cut the heads. There is little done to the heads after they are cut—a little trimming and possibly a minor wash to get the mud off. Packed in the field, the lettuce remains in those boxes through the process of cooling and shipping, until the grocer cleans it up and puts it on the shelf.

Lettuce growing and marketing is a risky business. "The lettuce industry is a hell of a crapshoot," Vic explains. "You might be better off in Vegas, where you're at least in a climate-controlled room while you're throwing your money away." The price of lettuce can change dramatically in as little as four hours. It's got to be available and it's got to be bought or it's lost.

"Yesterday lettuce was $10 to $11 a carton—that's twenty-four heads of lettuce," Vic said as we talked on a Saturday. "Today, it's $6 to $7. We've seen markets go from $11 to $4 in three days. To me the lettuce business is as pure an example of supply and demand as you can get.

"There's such a thing as a weather market, for example. The Salinas Valley in California has been shipping about 250,000 cartons of lettuce a day, and we're shipping 20,000 to 35,000 here in Colorado. If a weather system were to roll in off the coast to bring enough rain into the Salinas Valley to stop harvesting for two or three days, a freak situation that does occur, and if the price had been at the bottom of $4, it would go to $12 almost instantly. However, as soon as production started up there again, the price would drop back to $4. In the meantime we would have made a lot of money here. It happens.

"But it also happens—as it did this April,

Like most other vegetables, lettuce has labor-intensive harvesting. It is also a risky crop, with great price fluctuations. Most vegetable crops, including lettuce, do not have the government controls or guarantees that support many other crops.

May, and June—that lettuce sells just above the cost of harvesting, which is presently a little above $3 a carton. The grower has something more than $1,000 an acre invested in growing, and he gets nothing from it. That's pretty big money to lose.

"Fortunately, the opposite's true too. Take $10 lettuce, and your harvest costs and growing costs are still the same as $4 lettuce, and you've made a pile of money. I figure we sell iceberg lettuce below breakeven about 70 percent of the time, but we make it up when prices go up."

There may be money to be made in lettuce if you enjoy breathtaking crap games coupled with the tender loving care needed for your crop. There are enough farmers willing to live this way that we can have lettuce on the supermarket shelves every day of the year.

Artichokes

Artichokes are an example of a highly specialized crop that has found its niche on the American table. In total, artichoke production takes up about 12,000 acres of American

The United States doesn't grow as many artichokes as some countries, but it certainly has the largest single artichoke, standing at the entrance to a market and restaurant in Castroville. Practically all of the American crop of artichokes is grown within 15 miles of Castroville. Our production, however, is only 5 percent of Italy's total.

220

cropland. I know plenty of individual wheat producers who harvest more acreage than that by themselves. Specialty crops usually involve small acreage and a limited number of growers. They have to be marketed through regular channels that handle other, higher volume, fruits and vegetables to reach the final consumer. Some specialties are available only as fresh crops; others as processed. Some, like artichokes, are available as both.

In the United States artichokes are grown almost entirely within a few miles of the small California town of Castroville, where a sign over the main street proclaims it to be the Artichoke Capital of the World. Since Italy's artichoke production is over twenty times as much as ours and Spain's is ten times greater, this proclamation may be a bit presumptuous of Castroville, but then modesty has never been California's strong point. The state must have at least a dozen agricultural towns claiming to be the world capital of something or other.

There are fewer than fifty families involved in artichoke production around Castroville, but the production acreage has increased from 10,000 to 12,000 acres in the past ten years, and the average yield per acre has gone from 200 cartons to 300.

The artichoke is really a thistle, and what you eat is actually an immature flower head. (Given a chance to mature, the flowers are a lovely sight.) The artichoke is a perennial plant that originated in the Mediterranean area and is usually propagated by planting root sections from existing fields. Like potatoes, seeds can be planted but tend to produce plants that vary in size and quality; veg-

etative propagation is more dependable and quicker. Once established, the plants begin to bear about five months after planting and continue to be productive for five to seven years, until there is so much woody growth that production decreases. The plants are spaced widely apart, 8 by 9 feet being common, but as they grow they fill the interim spaces. The foliage of an artichoke plant resembles a thick fern.

Although almost three-quarters of the crop is harvested between March and mid-May, fresh American artichokes are actually available year-round. As is true for many other crops, such as oranges, the challenge and extra profit lie in figuring out ways to extend the season, to sell when others can't, and to get higher prices because of the decreased supply. One method used with artichokes is to cut the plants back below ground level between mid-April and mid-June to give a maximum crop in the fall or early spring, well before the traditional crop comes on. If the plants are cut back in late August or September, peak harvest can be late spring and summer, making a late-season product.

About 25 percent of the American crop is sold processed, either marinated in oil or packed in brine. Artichoke hearts are actually tiny artichokes, packed with the tough outer leaves removed, the spiny inner part too young yet to be fuzzy. Crowns, the delicious base sections without the leaves and fuzz, are also packed. A small number of fresh artichokes are imported from Chile and Ecuador—their peak season being the opposite of that in Castroville. Great amounts of processed artichokes are imported, with Spain being the largest source.

Like potatoes, artichokes are grown vegetatively—usually by planting root sections. The plants begin to bear about five months after planting and last more than five years (left). The edible part of the artichoke is actually an immature flower head, which, if allowed to mature, becomes a beautiful flower (right).

Although Americans are eating artichokes in greater numbers these days and enjoying them more, I doubt they'll replace potatoes—or even squash—as an American staple. Still, it's hard to say. Does Idaho have a 16-foot-high reproduction of a potato outside a restaurant featuring dozens of different dishes emphasizing potatoes? Castroville has such a structure in the shape of an artichoke, and it's been a successful attention-getter for more than twenty years, enticing travelers to visit a restaurant specializing in artichoke dishes. I recommend their French-fried artichokes or the artichoke omelet.

Fruits and Nuts

Fruits and nuts grown in the United States are almost as varied as the vegetables we can buy. Many of them are native to much of the rest of the world also. Walnuts, for example, grow in southern Europe, Asia, and the Caribbean, as well as both North and South America. River bottomlands in the central United States were once covered with black walnut trees. Oranges probably originated in the Malay Archipelago and migrated to the Mediterranean, then supposedly were brought to the West Indies by Columbus in 1493 and finally to the United States mainland. Peaches probably began in China, migrated to the Mediterranean, and were brought to Mexico by the Spanish before 1600.

Fruit and nut culture is distinguished from most vegetable growing by the length of time it takes to establish the plants and by the part of the plant that is edible. Whereas we eat many different parts of vegetables, we mostly eat the parts surrounding the seeds or the seeds themselves of fruits and nuts.

The fruit grower must plan carefully, since his planting investment is almost permanent. In many parts of the country the site selected to grow fruits is vital because microclimates can mean success in one location, failure only a short distance away. Cherries or peaches, for example, may do well on a hillside where air currents dump frost but fail at the bottom of the hill where the frost settles. Orange growers have often watched helplessly as their trees were killed by cold.

The advent of irrigation has made the entire California fruit and nut industry possible and has extended growing areas where rainfall is almost adequate but not dependable. In many sections of the country, irrigation is

Fruit is one of our great delicacies and is raised throughout the country, from Red Delicious apples grown in Pennsylvania (left), to peaches raised in the irrigated desert of Arizona (right).

Although California dominates in the production of many kinds of fruit, such as lemons, other states outproduce California in special fruits. Michigan produces five times as many cherries, for example, as California.

not entirely necessary but insures against a dry year or against drought-caused low quality. Although hillside planting has been favored as a method of avoiding frost, especially because the land is not suitable for row crops, this land was difficult to develop into irrigated country when flood irrigation, the flooding of entire areas, was the common technique. Drip irrigation, introduced a few years ago, made hillside growing less expensive and environmentally safer, with less possibility of incurring erosion.

Most commercial fruit trees are grown from grafted stock, with the rootstock being some hardy variety chosen to give the tree some special characteristics, and the upper part, called the scion, being a standard variety that has excellent eating qualities. Popular apple varieties like Red Delicious are grafted onto rootstock specially developed to

Flowers

Farm crops aren't really raised for the flowers they produce, but the flowers are the distinctive feature of the crop. While it may be less important for us to notice them, it is vital that insects do, for many of the crops are largely pollinated by insects.

Cotton (far left) could almost be cultivated for the beauty of its flowers, but corn has flowers (above) that are hardly noticeable. The horn-shaped flowers of tobacco (left) are responsible for the development of seeds so small that 20,000 seeds weigh only an ounce.

The blossoms of fruits such as the Bartlett pear (bottom left) are a familiar sight to many people in the spring, but how many of us have ever looked at a tomato bloom (bottom right)? Even fewer have ever seen okra in flower (near right).

The tiny flowers on a peanut plant (far right) are in the foliage above ground.

the frozen concentrated orange juice that appears on most of our breakfast tables is a relatively new addition to the citrus menu. It wasn't until nearly the end of World War II that there was any research on frozen concentrates, and it was 1948 before a basic patent was issued, covering much the same process that is still in use today.

The American orange groves that supply this juice are mostly in Florida. More than 90 percent of Florida's orange crop is grown for processing, whereas 80 percent of California's orange crop is for the fresh market, and only 20 percent is processed. While Florida dominates processing fruit, and California produces mostly fresh fruit, Arizona and California grow most of the lemons and Florida grows most of the grapefruit. Florida produces about 7 million tons of citrus annually, compared to California's total production of about 3 million tons.

Like most fruit and nut orchards, an orange grove is a lifetime investment. Orange trees a hundred years old are still producing, although newer varieties frequently replace aging trees and many groves, especially in Florida, have been replaced by urban development. In the fifteen years from 1970 to 1985, Florida's citrus acreage dropped more than one-third as its soaring population gobbled up the citrus land. (In forty years Florida's population has gone from 2 million to 10 million!) Citrus acreage in California has also been declining since the late 1960s, but not so precipitously as in Florida.

Commercial orange trees are grown from grafted stock, as are most fruits. In Florida lemon trees or bitter orange trees normally form the rootstock for oranges. One of the goals of grafting is to get trees into production more quickly. Oranges from seeds, besides having inferior fruit, often take fifteen to twenty years to start producing; grafted trees come into production in about five or six years. Rootstock is usually more hardy than the buds grafted to it, and sometimes during freezes the tree will freeze, but the rootstock will survive and new scions can be grafted to it.

The two varieties of oranges grown most are navels, which are the thick-skinned seedless oranges most commonly eaten fresh, and the thin-skinned Valencias, which have a few seeds, lots of juice, and are often eaten fresh but tend to be the standard for juice production. Almost equal numbers of navels and Valencias are grown.

Follow a navel orange from a tree to the market and you've watched a closely choreographed production. I drove through the groves around Orange Cove, California, with Jack Inman, who is the grower relations manager for Harding and Leggett, a packinghouse affiliated with the Sunkist Growers cooperative.

The greatest labor in harvesting occurs right at the start. Orange trees are large and oranges are picked by hand, each one judged for ripeness by the picker. Groves may be gone over several times in a season to select only the fruit that is ripe. Ladders have to be moved frequently, and each time a picking bag is filled, the picker has to climb down the ladder, dump the bag into a bin, and climb back up. It's hard work.

Inside the Harding and Leggett plant, the marketing process speeds up through automation and computerization. The plant uses

Oranges still require great numbers of pickers who have to judge each orange for size, ripeness, and quality. Many of the laborers for fruit and vegetable harvesting come from Mexico and travel from the Mexican border to Canada as the fruit and vegetable crops ripen.

six electromechanical graders developed by Sunkist engineers, each of which can grade ten oranges a second. Computers routinely produce records of fruit quality, size, and production from each of the grower's "blocks" of trees. The information is vital to the grower, who can see how he compares with other producers and how he can better manage his crop.

The contrast between the sleek, automated interior of the packing plant and the plodding climbing of ladders and hand selecting of fruit in the grove is startling. Although Sunkist engineers keep working to improve all phases of growing and harvesting, to date no one has been able to invent a field machine capable of selecting ripe fruit, picking it, and moving it to a collecting bin without damage.

Sunkist is certainly one of America's most-recognized household names. It belongs to

Sunkist Growers, a cooperative owned by the 6,000 citrus growers whose products it markets. Its annual sales run about $800 million, and its producers represent 60 percent of the total citrus growers in Arizona and California.

Cooperatives in agriculture tend to be huge organizations. Supply cooperatives—which sell supplies to farmers—and marketing cooperatives—which market farmers' crops—are both a major part of agribusiness. Total sales by cooperative organizations, of which there are almost 6,000, come to more than $60 billion annually, and there are eleven that do more than $1 billion each.

An average season sees Sunkist marketing some 68 million cartons of fresh citrus (each carton is 40 pounds), of which 30 percent is sold abroad, particularly to Asia. In addition to its sales of fresh fruit, the cooperative processes 800,000 tons of fruit into juices and peel products.

Orange groves require not only patience but a heavy investment of capital. Groves in both California and Florida have frequently been developed by investors who, until recently, have been able to deduct their investment costs from their taxes before there was any income from the grove. The new tax laws have made this less possible and thus made investing in new groves less attractive, which is probably one reason we are seeing the acreage in citrus diminish.

Freezing is clearly another reason for decreasing acreage. Citrus is a warm-climate crop, and there is very little agricultural acreage in the United States that doesn't occasionally have weather cold enough to cause freeze damage. The long history of heavy freezes in citrus continues to be devastating, even with all our agricultural advances. Florida, Texas, and California all experience freeze problems, although California seems to have the fewest.

When freezes threaten, radio stations broadcast urgent warnings, and growers go into action to save their crop. Many groves have towers sporting propellors with huge motors, which are activated to blow in warmer upper air. Smudge pots sit sullenly along the edges of groves, ready to be ignited. Even helicopters, which produce a heavy downdraft from their rotors, are pressed into service to force warmer air down to the trees. Growers tell me there is nothing more frustrating than to do everything possible to combat a freeze and then watch as the temperature continues to drop.

Freezes are by no means a new problem. Nearly a hundred years ago Florida's production had reached 5 million boxes annually when the state was hit with a massive freeze. It was fifteen years before Florida again produced as much as 5 million boxes. As a result of this devastating freeze, citrus areas shifted toward the southern part of the state, where freezing is less common.

In December of 1983, Texas was almost knocked out of citrus by a massive freeze. At the time Texas had considerable acreage in orange groves and was second only to Florida in producing grapefruit. Prior to the freeze, citrus production in Texas ran more than 650,000 tons annually. By the following season, production was zero and by the 1986–87 season had recovered to only slightly more than 100,000 tons. Even the replanting of the groves has been hampered because the existing nursery stock in Texas had frozen, and imports of stock from Florida have been banned because of a disease problem there.

Freezes and population pressure are not the only threat to the United States' citrus industry; imports are another. Florida's hold on the juice market is being eroded by the immense increase in production from Brazil. The largest-producing state in Brazil, São Paulo, increased its orange production from 1.6 million metric tons in 1968–69 to 10.2 million metric tons in 1984–85. Most of this production goes into frozen concentrate, the United States being the largest importer. In the ten years prior to 1985–86, Florida's production of frozen concentrate dropped by half, while imports from Brazil doubled during the same period, reaching twice the volume of Florida's production.

Frozen orange juice concentrate is a massive industry with standardized processing techniques. The oranges are trucked into the processing plant, selected and cleaned, then cut and juiced. The pulp and peel get shunted off to be dried for citrus pulp, which is mostly sold for animal feed. The juice, placed in a vacuum, goes through a series of evaporators that concentrate it into a syrupy liquid. Fresh

Citrus groves are long-term investments, especially if they are in parts of the country that require irrigation. Once established, however, production from groves goes on almost indefinitely.

237

juice is then added to give the concentrate the fresh flavor we like. The concentrate is then frozen and is maintained at 0 degrees Fahrenheit.

The marketing of processed citrus, including juices and other processed products, is highly competitive because there are no price or production controls and no control on imports. In 1985, when Brazil overproduced concentrate and dumped it on the world market, prices here dropped by 50 percent.

Fresh fruit, on the other hand, is subject to market controls. California's production of processed fruit, which is only about 20 percent of its crop, is mostly from fruit that has been processed as a result of the overproduction from its fresh fruit marketing agreements.

California and Arizona elected to operate with federally approved "marketing orders" for fresh citrus. (There are somewhat similar orders for many other crops, such as tobacco and milk.) Under the order representatives of producers and of the federal government sit down to decide how many oranges should be allowed to reach the market as fresh fruit during the main growing season. Each grower is then allocated his share of the production pie, and arrangements are made for new producers to be permitted to share in the pie. In effect this allows the states' production during most of the growing season to be limited to what they believe the market can absorb, the production being split among all producers on a historic production basis. With fresh fruit, there are volume limitations and quality and size controls, but no price controls. The roundabout purpose of this arrangement is to keep prices at a profitable level while maintaining a stable market.

Any excess fruit produced under the controls during the time they are in effect is "juiced," and juice is not controlled, except for quality. Exports of fresh fruit from California to outside the United States are not limited by volume controls, nor is any of the citrus production of Florida, either fresh or processed. The processed market, including imported juice, is thus relatively free, while marketing of most fresh fruit is controlled. The system is antithetical to our traditional free marketing but has been successful in lessening the volatility of the markets. Whether it has consistently given the consumer fresh fruit at the lowest price may be debatable.

The marketing agreements that are a common occurrence in agriculture would quickly run afoul of restraint-of-trade laws outside the agricultural field. Can you imagine automobile manufacturers, for example, being allowed to sit down and say: "Well, let's allow half a million cars to be marketed next week and decide how to divide them among us"?

Grapes

Grape growers are like lion tamers: their success depends on training the beasts! Left to themselves, grapevines grow wildly, sometimes as long as 50 feet. Pruning is therefore one of the major chores of maintaining a vineyard and usually results in cutting off about 90 percent of each year's growth. The goal of pruning is to encourage bearing and to keep the vines to a shape and length that makes picking convenient.

There are dozens of ways to train vines. Grapevines in the United States are trained to grow along horizontal wires supported on stakes that are usually less than 6 feet high.

The popularity of grape varieties changes, but most grapes in California vineyards are, like these Chancellor French hybrids, the result of hybrids that originated in France. Many of the French varieties are grafted onto stock from native American grapes.

The way in which grapevines are trained on wires varies greatly. They must be carefully pruned so they grow well but the grapes remain readily available for picking. Often 90 percent of the new growth is pruned away.

There are generally two or more wires, with the top wires supporting the part of the vine that grows foliage, the lower wire or wires generally supporting the canes that bear the fruit. One of the major costs of starting a vineyard is setting up the stakes and wires needed to train the vines.

Commercial vineyards are usually established from young plants rather than from seeds, since seeds frequently develop plants that vary greatly from the parent plants. Some grape varieties are grown from young plants rooted from cuttings of mature plants, the cuttings having developed roots in nurseries for a year before being planted. Other varieties are produced from cuttings that are grafted to a rootstock adapted to local soils or resistant to a plant louse called grape phylloxera or to nematodes or other diseases.

Different grape varieties have different sex configurations. Although some species of grapes have the male flowers on one plant and female flowers on another, most cultivated varieties are hermaphroditic—they have functioning male and female parts in the same flower. Most *vinifera* varieties, those most used in California, are hermaphroditic.

When breeders want to develop new grape varieties by crossing existing varieties, the procedure becomes very tedious. Using tiny forceps, the breeder must remove from the flower the male stamen, whose anthers produce the pollen. This leaves only the female part, whose stigma receives the pollen and whose ovaries are where the fruit develops. A paper or plastic bag is used to protect these emasculated flowers from receiving pollen from other flowers until pollen from the desired variety has been collected and can be introduced to fertilize them. The resulting

fruit and seeds will be a cross. The seeds are planted and after several years the quality and production of the plants are tested and the most successful vines selected. If the variety being developed is for wine grapes, the selection process is lengthened to include making wine from the test grapes and judging the quality of the wine.

Vineyard plantings vary in spacing according to variety, growing techniques, and harvesting needs, but rows are frequently 12 feet apart and plants are located about 8 feet apart in the rows. Once established, the spacing cannot be changed, but wires can be added or, if necessary, relocated, and vines can be trained differently. The fewer changes needed the better, however, because changes involve high labor costs and often cause short-term loss of production.

Once the vineyard is planted, commercial production is still as much as three to six years away, years during which weeds must be controlled, fertilizer must be added, pruning must be done, and stakes and wires must be maintained. Once in production, however, vines can be maintained almost indefinitely.

Grape growing and harvesting are slowly making their way into the mechanical age, but it's tough going. Fresh grapes and raisin grapes are still handled entirely by hand. Mechanical pickers—great lumbering machines that straddle the vineyard rows snatching bunches from the vines—are used in harvesting less than half of the wine grapes. They're a little like King Kong: too much muscle, too little brain. Engineers are constantly improving the machine's capabilities, but don't look for it to be harvesting your table grapes in the near future.

Watching crews handpicking fresh grapes

and packing them is seeing art in action as workers trim off unwanted grapes and arrange perfect bunches of grapes in boxes, pressured to maintain quality while also working quickly to get the grapes off to the packing houses so they can be rapidly cooled and loaded onto the trucks that always seem to be waiting. It's a hurry-up-but-be-perfect business.

Grapes have been cultivated almost as long as man has been cultivating. Hieroglyphics show how to raise grapes and make wine, and grapes, wine, and raisins are all mentioned in the Bible. *Vinifera*, which is thought to be the species mentioned in the Bible, still comprises more than three-fourths of the world's grapes if the hybrids derived from it are counted up.

Grapes were already growing in the United States when early settlers brought European grapes to the eastern United States. The native American grapes were resistant to native American pests, particularly a small louse, the grape phylloxera, but the newer European species were decimated by the louse. In fact, when American vines were shipped to Europe about 1860, they carried phylloxera, which destroyed about a third of the French vineyards within twenty-five years.

The best-known grapes in the eastern United States are still generally from native American stock. The American species are "slip-skin" grapes; the grape is eaten after slipping off the skin; the skin and pulp of European *vinifera* varieties are usually eaten together. Concord grapes, a variety of native American grape, are the basis of the New York State grape industry and represent about three-fourths of the American varieties grown.

These old vines have treelike trunks, and new growth is trained along the supporting wires. Vines last almost indefinitely if cared for, but the introduction of new varieties often necessitates tearing out old vineyards.

Facing page: New York's vineyards produce only about 3 percent as much tonnage of grapes as California's, and the mix of uses is different. The Concord, a native American grape, is the basis of New York's juice and wine industry. Two-thirds of New York's product goes to juice, whereas almost none of California's crop is made into juice.

Almost two-thirds of the New York crop goes for juice, one-third ends up as wine, and very little is sold as fresh grapes. Although New York is the second most important grower of grapes, it represents only about 3 percent of the tonnage produced by California. California's immense production is used differently from that of New York. About 14 percent is sold as fresh grapes, 30 percent as raisins, and 55 percent is processed for wine making.

Classifying types of grapes is a bit touchy, although the usual terms are *fresh* or *table*, *raisin*, and *wine*. In practice, however, both raisin and fresh grapes may also be used in wine making, in addition to the large

amounts of grapes grown specifically for wine. After the finest of the fresh grapes and raisin grapes have been removed from the vines, those same vineyards are usually gone over again for the remaining grapes, which go to wineries. Both fresh grapes and raisins demand physical perfection; wine grapes end up being crushed.

Of the many varieties of fresh grapes available, Flame Seedless and Flame Tokay are the most common. Thompson Seedless grapes are by far the most heavily grown raisin grapes; in fact, they occupy almost 40 percent of the total grape acreage and are used

not only for raisins but as fresh fruit and for wine. Seedless grapes, such as Thompson Seedless, are grapes whose flowers have been fertilized but in which the developing seeds spontaneously abort, leaving the grapes seedless. There are some other grape varieties that can produce fruit without fertilization.

Grapes grown in California for white wine represent more than one-fourth of the total grape acreage, the most common being French Colombard. Zinfandel is the predominant red wine grape, and red wine grapes represent about 20 percent of the total acreage. The color of the wine, incidentally,

Table grapes, packed and sold fresh, require great amounts of hand labor. Although mechanical pickers have made some headway in picking wine grapes, table grapes are completely hand picked. After picking, they must again be selected and trimmed to fit the packing boxes that end up in grocery stores.

244

is not determined by the color of the grape skin but by the color of the pulp. White wine varieties may have red pigment in the skin if the juice is colorless; grapes with red skins or red juice are used only for red wine.

Like other fruits and nuts, grapes are a crop requiring heavy capital investment. California grape acreage currently is estimated at about 700,000 acres, and if you're interested in buying land not already into grapes and converting it to a vineyard, look for land costs of $5,000 to $12,000 an acre, and planting and maintenance costs for three to six years after planting before there is appreciable income. Not for the light pocketbook.

There's a reverse side to the coin, of course. Once established, grapevines go on indefinitely. There are vineyards in California that have been in production for more than a hundred years. New varieties may take acreage from existing vineyards. There were only 90 acres of Flame Seedless grapes growing in 1975; by 1985 there were 18,000 acres. With 18,000 acres of new vineyards, there had to be about 9 million new plants produced within a ten-year period!

Advertising would have you believe that after science there is magic in wine making and grape growing. That may be true, but not any more so than in any other crop. There just happens to be more money available for promoting this image. The growing of corn or wheat just doesn't seem to convey the romance of grapes and wine. Corn is certainly more important to our lives than grapes, but there's no clinking of glasses or breathing in the aroma or tours of the corn country. How about more romance for wheat? To be honest, I'd rather ride a wheat-tasting train through the Palouse of Washington than a wine-tasting train through California's Napa Valley.

Kiwifruit

There are many vegetable and fruit crops that will never make it to the agricultural big leagues but are well known and often delicious. These crops are frequently successful through the efforts of one person or a few producers in some small area. Ordinarily their success or failure depends more on salesmanship than on agricultural performance. Most of these small crops are helped to success by having been adopted, and often adapted from, other parts of the world.

Kiwifruit is typical of this kind of crop. In the United States kiwifruit is grown entirely in California and occupies just a little more than 4,000 acres—less than 3 percent of California's entire citrus acreage. It is promoted by the California Kiwifruit Commission which, like other agricultural associations, is funded by a checkoff, an assessment on the fruit each grower produces. Their publicity blurb on kiwifruit begins:

From the deep, misty valley of the Chang Kiang in an era known only to the winds and sands of time, first sprang the Yang Tao. For centuries it clung to the soft, dry earth gaining sustenance and maturity from the warm spring rains, slaking its summer thirst from the mighty Yangtze, bearing its reward in the soft humidity of the fall, and to sleep undisturbed under the solid, star studded winter sky. Here in this ancient valley, nurtured with the residue of Tibet, and the dales of southeast China, developed the fuzzy, unattractive Yang Tao. Like the fables of old, one is not surprised to discover that ugliness, like beauty, is but skin deep.

Eventually the reader learns that it was nearly fifty years later that seeds of Yang Tao were first sent to England's Royal Botanical Gardens and not until 1937 that there was really any commercial planting in New Zealand. Further, it was another twenty years before the product was first exported to the United States. Only then did the final name change take place, and alas, the publicity blurb doesn't mention how this was managed. Yang Tao, which had become *Actinidia chenesis planch*, and also Chinese gooseberry, marvelously became kiwi. Once established as a crop in the United States, however, some nameless soul decided to differentiate New Zealand's kiwi from the United States' kiwi by naming the latter kiwifruit.

Doug Wilson is a kiwifruit grower in California whose acreage is near Gridley, north of Sacramento. Doug calls himself a city kid. "I had a business degree, and a degree in agriculture, and I visualized myself as a research scientist in biology or soil science, but I didn't really see myself as a practical farmer," he says. "I had never even driven a tractor."

He is a product of the 1960s, with a stint in Vietnam, and he has a relaxed way of talking. "I got a job with a kiwi nursery in 1972. That introduced me to the business. It was real easy to believe in it because of the qualities the fruit had; it was easy to convince people to plant the vines. You could start a nursery and within six months have enough orders for plants, with deposits paid on them, to be operating profitably; plants not to be delivered for two years.

"Then I managed a kiwi ranch down in the Turlock area. It was just at the beginning of the kiwi expansion period. I learned how to farm, really, over about four years. The whole operation was being very successful, the crop was coming into production, and the projections on production, prices, and profits all looked good. Using my experience there and the positive outlook for kiwis, I was able to talk someone into believing in me enough to buy this ranch I'm now on, so I came up here to run my own operation in 1978.

"When we bought the ranch, it was in freestone peaches, one of the last freestone ranches in this neighborhood. It wasn't that the farm was really unproductive; they had about five different varieties of peaches, harvested from early May through September. But it was really a one-man operation, a small family farm for peaches, and it couldn't have existed in today's agriculture. With peach prices as they are now, there just wouldn't be enough income to warrant growing peaches on this little place. Kiwifruit production can give a living from small acreages."

Production figures back up what Doug says but indicate that increasing competition takes its toll. In the 1980–1981 crop year, California's production was just shy of 8 million pounds; by the 1985–1986 year it was more than 34 million pounds. The average pound value in 1980–1981 was $1.62 a pound, but by 1985–1986 it had dropped to less than a dollar. The moral to the figures is to get in early.

This is the typical cycle for all new crops. While the crop is still being produced on a small scale and being heavily promoted,

The vines from two rows of kiwifruit plants are trained to an archway made of wire, so that pickers seem to work in a tunnel, reaching up to pick hairy brown eggs. The fruit can be kept in cold storage for as long as six months. California's production, harvested in October, competes little with the fruit imported from New Zealand, since their production seasons are reversed.

prices are high and profits are great. The heady profits attract capital, making expansion easy. The markets grow readily. Eventually the expansion of production reaches its limits and the market becomes saturated, resulting in greater competition and lower prices. Economic theory reduced to kiwifruit vines.

Kiwifruit actually grows on a vine, and its cultivation resembles grape cultivation. The plantings in the United States are called vineyards, although in New Zealand they are referred to as orchards.

Like grape vineyards, kiwi vineyards take years to establish. The first year the grower plants bare rootstock, and the growth from that gets trained up onto a trellis; the second year the "fruiting canes," branches that will bear fruit, are established. By the third year there's a small amount of fruit, but it's not until the sixth or seventh year that production reaches major volume.

Unlike grapes, which require massive amounts of training on wires, pruning, and managing, kiwi vines take relatively little thinning, and the vines seem to last indefinitely. So far, insects have been a minor problem, with spraying generally unnecessary. But as is typical of new crops cultivated in increasingly concentrated numbers, insects are bound to build up significantly.

The plants flower in May, and the fruit matures by the middle of October. The vines grow into a canopy covering the space between two rows. The canopy is most readily described as being a lengthy 7-foot-high cavern with dark green walls and hairy brown eggs hanging from the top.

"Picking is done mostly with local labor," Doug Wilson explains, "and my help gets paid on an hourly basis, rather than by the number picked. Since the culls are valueless, it takes a keen eye to make the decisions on which fruit gets picked, and it isn't like walnuts or pecans; the fruit has to be treated with care."

One of the problems of kiwifruit is that so far there are few by-products. Only the very top-quality fruit can be marketed; the rest are discarded. Unlike orange growers, who can "juice" the imperfect fruit, or the growers of many other crops whose imperfect fruit can be diced or sliced or squeezed into cans, the kiwifruit producer can't make money from culls. American ingenuity, and the big

dollars involved, will bring a remedy to this. Look for kiwi juice or kiwi jam or kiwi topping or frozen kiwi pieces to appear on your local grocer's shelf. The kiwifruit publicist insists that even the skin can be used; it contains an enzyme that makes an excellent meat tenderizer.

Although kiwifruit has few by-products, it has two great marketing advantages. First, the fruit can be stored—at 31 to 32 degrees Fahrenheit in 90 to 95 percent humidity—for six months. The producer can thus keep the product supplied over a long period of time even though all the kiwis are harvested in a few weeks in October, and the consumer can keep the precious product a long time in a home refrigerator.

Second, unlike any other crop I know of, the imported kiwis from New Zealand compete with the California crop very little, for New Zealand ships in during our summer months and the California crop is sold during the winter months. New Zealand kiwi are thus competitive with summer fruits like melons, peaches, nectarines, and plums. California kiwifruit competes with apples, oranges, and bananas. This has led to cooperative kiwi promotions between New Zealand and California, and to the immense marketing advantage of having kiwifruit available year-round. New Zealand production runs about five times that of California, although the gap is narrowing.

International production of kiwifruit is increasing rapidly. Growing areas of the world are limited to climates that have hot summers and cool winters like California, which presently exports about half of its crop. Production is coming on in Australia, Chile, Corsica, France, India, Italy, Japan, Korea, Pakistan, Russia, South Africa, and Spain. China, where it all originated, has not yet gone into the business. Kiwi may be too exotic a product for the proletariat, at least for the moment, or perhaps the needed refrigerated storage isn't yet available. But China is surely bound to join the producing throngs eventually.

Lest you feel kiwifruit is just one more food fad whose time has come and will go, don't underestimate the ability of the promoters:

> All this history would be meaningless if
> it weren't for just one thing . . . the fuzzy,
> somewhat nondescript Kiwifruit tastes
> wonderful. Its merits have been 'discovered'
> by everyone from the Nouvelle Cuisine
> in France to the fine restaurants of six
> continents, and by consumers shopping in
> their favorite market throughout the world.
> The rich, emerald green meat of Kiwifruit
> is more than something to attract the eye as
> a complement to fine food. Its taste has been
> compared to that of strawberry, banana,
> grapes, melons, pears, peaches, and a host
> of other delectable fruit combinations.

Almonds

Americans eat more almonds than any other nut and, fortunately for California's almond growers, so does the rest of the world. That hasn't happened by chance but is the result of sales efforts, mostly by the California Almond Growers Exchange, the almond cooperative with worldwide sales of almost half a billion dollars. In recent years foreign sales have also been helped by the weaker dollar that provides a better almond bargain to the rest of the world. The entire American crop, usually more than half the world's supply, is grown in California, and almost three-fourths of our almond production is sold abroad. More than 5,000 growers market their almonds under the Blue Diamond ® label that belongs to the California Almond Growers Exchange.

Almond growers watch the sky carefully late in February or early in March when their crop is coming into bloom, not just because the almond bloom can be damaged by rain, but also because the bees that pollinate almonds require as much time as possible to work on the flowers, and wet weather hinders them in their efforts.

Although bees are needed for all fruit pollination, almonds present special problems, because almond flowers, beautiful as they are, don't produce much nectar. Apiarists, the people who own bees and hives, rent their bees to crop growers whose plants need pollination, moving the bees from one orchard to another as new crops come into bloom. Apiarists earn income from the rental of bees to a limited extent, but primarily from the honey produced from the nectar. The rental of bees is therefore more costly for the almond grower than for the orange grower, for example. Orange blossoms have a sweet fragrance that attracts the bees, and they produce large amounts of the nectar that is made into honey. Almond blossoms, which have little fragrance, require about three beehives for each acre of trees for adequate pollination and produce very little nectar.

Worry about weather in many almond-growing areas of California extends to frost problems also, for almonds are a warm-weather crop sensitive to cold. The almond originated in southwest Asia, and most of the foreign growers are in the warmth of the Mediterranean area, Spain being second to the United States in production. When a frost warning is out, California growers who have sprinklers or who irrigate by flooding their groves turn on the water to prevent the bloom from freezing. One grower told me that when frost is a danger, he floods his grove with water from his wells rather than from the irrigation ditch; the ditch water comes directly from snow melt in the Sierras and is likely to be less than 45 degrees, while the well water is above 60 degrees. The rest of the year he uses the ditch water because it is cheaper than the water he has to pump with electricity.

Primarily because of sensitivity to storms, the annual production of almonds varies greatly, and prices vary with the quantity of production. In 1983 a storm at the wrong time cut production to 242 million pounds, but in 1984 there were no weather problems and production more than doubled, to 590 million pounds. Fortunately for the industry, almonds are like wheat or corn; they can be carried over in carefully controlled storage from one year to the next, and the reserves are used to smooth out the peaks and valleys of production. Almond sales are extremely price sensitive, since almonds are a crop buyers can live without. The growers hate to see their market hurt by high prices caused by shortages, especially when the shortages are followed by overproduction in a good year. They want the buyers to form "the almond habit," and high prices disrupt their planning. (On the other hand, of course, the sales promoters don't want prices to be so low the producers they represent operate at a loss.)

The almond acreage in California has more than quadrupled since 1950, but the growth seems to be leveling off now, with new plantings of less than 3,000 acres a year in recent years, down from more than 20,000 acres a year in the early 1980s. Once established, almond trees produce almost indefinitely. More than 400,000 acres are in almonds, over twice the California orange acreage. Although the prices paid to growers for almonds have generally gone up very little in the past fifteen years, profitability has been helped by increased production per acre, and harvesting costs have dropped greatly as a result of mechanization.

The almond most closely resembles the peach, and occasionally almonds and peaches hybridize. The nuts grow thickly on the trees. They have a heavy leathery outer covering, the hull. As the nut matures, the hull dries out, splits open, and curls outward, but the nut in its shell remains attached to the hull and the hull to the tree. Before the 1960s almonds were picked laboriously by hand. Today harvesting is mechanized, and almonds, still in their hulls, are now shaken from the tree by a hydraulically powered shaker mounted on a tractor. Within a few seconds the entire tree's crop is shaken off and lying on the ground. Mechanical blowers, also mounted on tractors, blow the nuts into windrows from which they can easily be loaded into trucks.

Like other farm producers who have found value in by-products, almond growers separate the hulls from the almonds and sell the hulls as cattle feed—about 200,000 tons a

Mature almonds have a tough outer hull that covers the thin shell of the nut. As the nut matures, the hull dries out and cracks open, but the nut and shell remain attached to the hull, which remains attached to the tree.

Until sometime in the 1960s, almonds were harvested by hand, but now a tractor-mounted shaker vibrates the trees, and after a few seconds the entire crop drops to the ground. A sweeper then moves the nuts, still in their hulls, into rows; they are lifted into trucks with automatic equipment, then separated from the hulls.

year. The almonds themselves, still in their shells, go on to the processor. The California Almond Growers Exchange shells so many almonds each year that, as a profitable by-product, the shells are burned in an electric generating plant that produces enough electricity for a city of 10,000 people.

Almond consumption by each of us in the United States has doubled in the last fifteen years, although we still average less than three-fourths of a pound per capita. Although almond growers are looking forward to increasing that to a pound per person through adroit promotion, their biggest sales effort has been toward increasing foreign sales. In the five years from 1980 to 1985, the world consumption of almonds grew from 450 million shelled pounds to 750 million shelled pounds. West Germany is the largest importer of American almonds. The Soviet Union has occasionally been the second largest, but as is true of other crop imports there, the market varies according to political whims, not consumer desires. Japan has been a major marketing target; its almond consumption has risen 150 percent in five years. In Japan there are more than a hundred almond products available, including dried baby sardines served with slivered almonds. The product has been so successful, in fact, that the supply of sardines hasn't been adequate to keep up with the demand.

The Changing Scene

American agriculture still has the image of an unchanging entity fixed in time. The rest of the world seems to be evolving while dependable old agriculture remains unmovable. This notion is usually accompanied by a picture of the farmer as an uneducated hayseed who cannot or will not keep up with the urban world. That simply isn't the way it is. Agriculture is dynamic and changing.

Agricultural change comes about through several avenues. First there are technical changes—improved animals, crops, equipment, techniques, chemicals. These are the result of research, however that might be defined. Then there are economic changes necessitated by market changes. Recently these are largely the result of overproduction, but they also include market acceptance or rejection of products. Finally, there are politically originated changes. These are actions by the government to correct problems caused by the research and marketing changes.

Agricultural research is active in a number of places. The massive bureaucracies of the land-grant universities and the Department of Agriculture are engaged in research, as are private companies looking for products that will sell better and innovative individuals whose independent research breaks new ground. Government and private industry often cooperate, with government research ordinarily aimed at the more basic problems and private companies working on problems closer to the market. Most of us recognize that farming has long had more government assistance than other industries, and government-funded agricultural research is far more extensive than that carried on for other fields, with the possible exception of the health field. The research bureaucracy has had a long time to build up and, in its ponderous way, is effective at changing agriculture. Farm research by private companies is inclined to move more rapidly.

Another source of research is simply the individual farmer, whose curiosity and ingenuity have led to many innovations which are originated by farmers themselves. While in the past farmer inventions tended to involve the mechanics of new or improved machinery, as farm education has expanded, farmer contributions to new ideas are increasingly in animal or crop improvement. Better farm recordkeeping is, by itself, a major factor in research; knowing what is actually taking place is one part of doing something new.

Economic changes are primarily the result

Agriculture changes at least as fast as the rest of America. Its changes depend both on new technology and the availability of capital.

of successful research, not only in farming but in processing and use of farm products. Generally, we have been far too successful in enhancing production but not sufficiently successful in increasing consumption. If we are not well fed, it is mostly because we choose the wrong foods, not because the foods are unavailable.

The economic problems of agriculture are largely the result of low prices for farm products caused by a surplus in both the domestic and foreign markets. The overproduction that causes this has been possible because of increased know-how, more efficient plants, and chemicals and machinery replacing labor, allowing for increased productivity and also for increased efficiency with larger production units. All of this has raised the need for agricultural capital, which until recently has been readily available. I simply can't believe these technical changes will cease or even slow down. Pundits of agricultural doom have been predicting decreased productivity for many years, but this seems unlikely, given adequate capital.

The future continues to lie with controlling agriculture to keep increased productivity in line with market demand. This can only be done through government controls, without which there would be an upheaval so far-reaching that it would most likely cause a general depression of mind-boggling size.

New Crops

Hovering in the wings, waiting to enter agriculture's main arena, are any number of promising new crops. Many will never make it, and a few will manage a tentative entrance. At the least, some will use up acreage, providing a product previously unavailable. At the most, one will become a new star. These potential new crops are an example of change occurring in agriculture, and similar chapters could be written about new or improved farming techniques, improvements in existing crops and animals, or new chemicals. They all represent change.

The best possibility for a new crop being accepted is if it has the potential for major industrial use, rather than as food. Existing crops have already entered the industrial market, especially corn, but their use in industry has had relatively little exploration. As we exhaust our nonrenewable resources, crops will become more important for their industrial applications.

Some new crops are being extensively researched by government agencies, although the government is inclined to work on improving existing crops. A handful are subject to development in backyard plots with amateurs determining the progress, and a very few are being developed privately, spurred on by adequate venture capital. There are always some that are just promotional schemes that will disappear when the spotlight is turned on them. It's a tribute to American enthusiasm that the potential to develop new crops exists.

There are some 200,000 species of flowering plants known to man, and although several thousand have been used for food, only about 200 species have actually been cultivated. American agricultural history is filled with examples of plants that have been known about for many years but have had no commercial development. Two major crops today are good examples: soybeans and peanuts. It was well into the twentieth century before either was commercially developed in the United States. Although our knowledge of little-known crops has increased, there are unquestionably many presently unused curiosities that eventually will become important to us.

Where will these plants come from? Some have always been around but have never before been developed—jojoba is an example. Others have been grown abroad but could be grown here—quinoa and kenaf are examples. Genetic manipulation will contribute to the pool of crops that can be developed, but it is

This government scientist experimenting with tissue culture may be working on completely new ideas of how crops can be propagated and grown.

still early to understand how much of an effect this may have. My guess is that genetic manipulation will eventually have immense effect on the quality and characteristics of existing crops but for some time it will not be used to develop totally new crops.

Developing crops commonly, though not universally, go through a standard cycle. A crop that seems already to have a limited market but even more limited production catches the attention of a promotion-minded person or group who announces the discovery of an agricultural dream: an easily grown crop that brings incredible prices. Investors, large or small, rush to put money into the crop by buying planted land or buying seed at a very high price or simply by paying for research or development. Usually very little is done to investigate or quantify the existing or potential market. For most people, growing a crop is more exciting than marketing it.

Even with the crop not yet well developed or in high production, as soon as increased production reaches the existing market the market is overwhelmed, the price drops, and the bubble is burst. Sometimes there are investors who can afford to hang on, taking losses while they develop a better product and find ways to expand the market. More likely the promoters go on to some other pie in the sky.

Three examples of new crops—jojoba, quinoa, and kenaf—reveal very different aspects of the same development.

Jojoba

Jojoba is a native American plant that grows in the Sonoran Desert. Its beans, which are about 50 percent oil, have long been picked by hand by Indians to supply a small market. Very high in quality, the oil has been in demand for cosmetics and in some cases has been used to replace sperm oil, which originally came from sperm whales, now an endangered species.

In the Casa Grande area of Arizona in the early 1980s, picking wild jojoba beans became a highly competitive but profitable part-time activity, not only for the local Indians but for bored retirees. Seeing the existing market for the wild beans, developers set up companies to sell desert land to investors and, for a fee, to plant the land to jojoba. Tax laws being as they were, investors could deduct their expenses against other income until the plants began to produce in three to six years. Prices of jojoba beans soared even higher as growers added to market demand by buying beans to use as seed for the expanding plantings. Magazine articles from 1980 to 1985 touted jojoba with such titles as "Taming the Wild Jojoba," "Jojoba Sounds Unique," and "Jojoba: The Oilseed Word for Profits?"

But nothing is ever as simple as it seems. The plants froze easily, the male to female ratio turned out to be all wrong—only female plants bear beans—and in 1985 the Internal Revenue Service decided that investors really couldn't take all those deductions. The project collapsed.

Meanwhile, there were other jojoba growers in western Arizona and California. They also had difficulties, but they were better funded and were solving problems. They hung on. Although the acreage planted to jojoba diminished for a while, gradually the plants began to produce. The total production in 1986 came to about 1.5 million pounds of beans. There was a market for the beans, and the growers are belatedly investing money in research to develop new markets. In the fall of 1987 the first Jojoba Harvest Festival was held in McVay, Arizona. The turning point, proponents say, has come. The outlook for the future remains to be seen.

Kenaf

Kenaf is another newcomer, an East Indian hibiscus that has been used in the past as a substitute for jute in making sacks. Never grown in the United States, it has been researched for potential development here by newspaper publishers looking for a cheaper source of pulp for newsprint.

An annual plant that grows 12 feet high, kenaf produces lengthy fibers, can be grown as a row crop, and is easily harvested mechanically. Since it is basically tropical, the best place in the United States to grow it is in southwest Texas, along the Mexican border. If a need for large acreage develops, however, it is likely that it will be grown in Central America.

Early pulp-making experiments indicate that kenaf produces excellent paper, mixes well with ordinary wood pulp, and does a superb job of improving the quality of recycled paper. It is too early to say whether it can replace wood pulp economically. Besides collecting statistics on production per acre, soil conditions needed, and insect damage possibilities, researchers are surveying the potential market. Early figures indicate it would take 40,000 growing acres to supply each of the sixty-five pulp mills that make newsprint, or a total of more than 2.5 million acres

planted to kenaf. That's a lot of farmland.

The development of kenaf has been well funded by venture capital raised by newspaper groups in the United States and Canada. Unlike many of the other developing crops, it hasn't had to rely on backyard tests and back-of-the-envelope figuring. By the time the experiments are completed, the research will be pretty conclusive as to whether the crop will be agriculturally and economically feasible as a source of paper pulp and whether its excellent fibers might be used in other products.

Quinoa

Quinoa (pronounced *keen-wah*) is something else again. A type of pigweed grown in the high Andes, it has seeds that are used as a kind of grain. Many people enjoy it cooked as a breakfast cereal. It has a nutty taste, and the small grains stay separate and crunch nicely. Since its introduction into the United States, it has done well in health food stores at very high prices, $3 to $4 a pound, and is presently being pushed toward national distribution.

To date the quinoa sold here has been imported from the Andean countries. "Patches" of quinoa are being grown in several spots in the mountains of Colorado, maybe 200 acres all told, spread over several hundred miles and in areas of varied climatic conditions, in an effort to determine its growing limits and to develop more dependable seed. Unlike the development of kenaf, which is financed by a joint venture of newspaper people who will provide the eventual market for it, quinoa's experimenting is being done partly by Colorado State University, partly by Sierra

Blanca Associates, a private research group, and partly by volunteers. It's pretty much a homespun effort; fun but a little decentralized, to say the least. Should its production be feasible, American quinoa would still have to be competitively priced with the Andean product.

There are dozens of other plants that might or might not find a home in the United States. The Department of Agriculture constantly looks at new possibilities and experiments on a small scale. Potential marketers push crops with varying degrees of realism, and social reformers who want to save the world through better nutrition get involved. Is there another soybean or another corn waiting to be discovered? I see none, but don't rule out the possibility.

Quinoa, a grain from the high Andes, has a nutty taste as a cooked breakfast cereal and is currently available in many stores at high prices. Grown only experimentally in the United States, mostly in the Colorado mountains, it will have a long period of development before it succeeds or fails, both as a crop and as a product in the market.

Alternative Agriculture

The term *alternative agriculture* has a fairly solid core, but it gets progressively fuzzier toward the edges. The core meaning is, simply enough, an agriculture different from the predominant agriculture we now have in the United States. The fuzzy part comes when we try to define what the advocates of alternative agriculture propose to make different.

Since there are many alternative movements at work, generalities are difficult to make. Some movements emphasize economics, politics, and sociology; others dwell on farming techniques. A look at the most common disagreements that alternative agriculture has with establishment agriculture usually starts with the argument that the family farm has been, or is being, wiped out, replaced by corporate farms. Alternative movements view the family farmer as the "backbone" of the country who must therefore be preserved at any cost.

They worry about off-farm input, advocating that the farmer be much more dependent on his own efforts and materials, less dependent on off-farm input—more physical labor, fewer mechanical labor savers; more manure, less purchased fertilizer; more home-baked bread, less store bought. They advocate greater crop and animal diversification. They are greatly worried about any kind of chemical used in farming, by which they mean purchased fertilizers, herbicides, and insecticides. Most of their arguments are, in effect, aimed at turning back the agricultural clock to the simpler earlier days.

They are usually insistent that their methods, whatever they are, will improve the condition of the soil and that present farming techniques are rapidly making the soil unusable. *Sustainable agriculture* is their most recent buzzword, meaning a way of farming by which the soil, and in fact the whole biosphere, will be kept in good shape.

Organic Gardening

Best known among the alternative movements is organic farming. Spearheading this movement is Rodale Press, a publishing empire founded by the late J. I. Rodale and now headed by his son, Robert. The organic movement swept into great popularity in the rebellious period of the 1960s, when anything the establishment did, including farming, was challenged. It does have a much longer history, however.

Essentially, organic gardening or farming rejects the use of nonorganic fertilizers and pesticides in favor of the use of organic materials such as sewage sludge, animal manures, and crop residues to improve the soil and provide needed chemicals to crops. It promotes the control of insects through a number of techniques, including the use of diversified cropping systems and the use of "good" insects to combat "bad" insects.

Rodale has endeavored, and largely succeeded, in moving from serving what I would call a relatively faddist readership for their original magazine, *Organic Gardening*, which teaches a specialized organic theory of backyard gardening, to reaching a wider general audience. The Rodale publishing business now includes a health magazine, *Prevention*; an extensive book publishing operation, Rodale Press; and several other magazines, including *The New Farm*, that is aimed toward promoting the use of organic techniques in commercial farming.

The Rodale empire also includes a research farm devoted to studying and promoting organic techniques and has extended its efforts to occasional testimony before Congress or other legislative bodies. It has even evolved a sort of philosophical group, the Cornucopia Project, aimed toward helping with many of the world's problems—primarily food production. It is currently also getting involved with community development.

Rodale has done a remarkably successful balancing act, interesting the little old lady in sneakers, with trowel in hand; the young couple trying to escape the urban blahs by putting their toes in the dirt; and all of us (including, certainly, me) who have some intuitive desire to grow things and to live, in some vague fashion, in greater harmony with nature.

It's my belief that the Rodale group has been less successful in reaching commercial farmers who are dependent for failure or success on the bottom line. Here organic theorists have had to contend with the reality of an entrenched agricultural bureaucracy, including the companies that manufacture the products Rodale wants to do away with; most of the academic agricultural establishment, which has generally been reluctant to accept organic theories or practices; and also the farmers themselves, who develop their farming plans with computers, through the universities, and through their accountants.

If Rodale has succeeded in any way, it is in making farmers and the agricultural establishment think about alternatives, especially in such areas as pesticide control. Although organic adherents haven't achieved their goal of doing away with insecticides, the brouhaha they have helped to raise is succeeding in getting the attention of the establishment, making them realize that something better must be done.

In keeping with my belief that, aside from the abrupt effects of legislation, agriculture moves in swells rather than choppy waves, compromise is having its effect. The dangers of insecticides are common knowledge now, and the ability to cut back on their use through more sophisticated farming techniques and through development of safer chemicals is generally being worked on. This is partly because of pressure from the organic movement, partly because of the increasing cost of chemicals, and partly because of competition among the chemical producers. So far however, I see almost no movement toward accepting the idea of no chemicals at all.

There are countless other organizations concerned about the plight of the family farm, the loss of farm ground to urban development, chemicals, damage to the land, animal welfare and animal rights (the former want animals treated better, the latter want them treated more or less as humans), vegetarianism, two to twenty acres for everyone and, if not independence, at least happiness.

Some of the organizations are set up to change public opinion through education and to change legislative minds. Some are set up to teach participants how to change their farming techniques or their "land philosophy." Some are set up to raise money to achieve their goals through land purchases. A few are set up just to raise money to pay salaries.

Traditionally, the Department of Agricul-ture has been oriented toward the commercial farmer, as have the agribusiness companies. However, pressure from alternative agriculture groups has at least forced the USDA to notice them. This has led the USDA to establish the Alternative Farming Systems Information Center at its National Agriculture Library and to set up an Office of Small-Scale Agriculture.

The universities have dragged their feet on being interested in organic and other alternative questions. They claim to be teaching what their research has been proving is the right way and say that when alternative agriculture shows it is worth developing, they will. Alternative agriculture groups usually charge that university research is under the control of the large agribusiness firms who pay for part of the research and whose business would generally be harmed by the adoption of alternative agricultural practices. The universities deny that. These days alternative agriculture does seem gradually to be getting more exposure in classrooms and research stations.

One of the more interesting offshoots of the alternative agriculture and "back to the earth" movements is the beginning of new schools established for that purpose. In Colorado the Malachite Small Farm School has put together a curriculum that includes bee-keeping, farming with horses, and the fundamentals of milking a cow. The studies are clearly not aimed toward making a living by farming (although some graduates have gone into small farming), but more toward creating a specific approach not only to farming but, in fact, to life. Its students are far more likely to be from the urban towers of New

Lettuce grown hydroponically is being produced commercially at Archer Daniels Midland in Illinois. Hydroponics is the science of growing crops in chemical solutions, without soil.

York or Denver than from the farm country of Iowa.

The American civilization is built on adversarial relationships that get resolved by compromise, and I believe that is the way we function best. We are excellent at putting opposing views in front of everyone, and in the end the action we take is usually a compromise, ideally extracting the best from the opposing arguments and moving forward. I hope this continues to work in agriculture.

Crops without Soil

Hydroponics—the growing of plants in nutrient solutions—is a form of alternative agriculture. Actually it's the opposite of organic gardening, which insists that soil is the basis of all growing. Hydroponics does away with the need for soil by applying the needed chemicals to plants directly, either in a water solution through the roots or through some base such as sand.

Hydroponics is likely here to stay, although it isn't about to revolutionize agriculture except in very limited, specific situations. The basics of hydroponics production of vegetable and horticultural products are well known, the chemistry is quite standard by now, and the mechanics function well. But hydroponics has had a hard time really getting under way in the United States, with one commercial failure after another.

One of the major agribusinesses, Archer Daniels Midland, has invested in a large hydroponics facility in Illinois. They call their operation "hydrofarming." In their 10-acre hydrofarm greenhouses they are presently producing 30,000 heads of lettuce every day. Lettuce is the major crop grown, but there are other crops such as herbs. ADM is bringing year-round vegetable production to the Midwest, decreasing the region's winter reliance on products shipped in from the Southwest and California.

Using sophisticated techniques, the greenhouses grow the lettuce in a water solution instead of soil. Seeds are put into "growth cubes" and carefully nurtured under constant illumination from grow lights. The young plants are moved.into a seedling area, then finally into the production area, where they mature. It takes only twenty-one days in the production facility to make a lettuce crop, possibly less than half the time if grown outdoors.

ADM is also a major developer of high-fructose corn sweetener. As part of that product's manufacture, excess heat is produced, which is ducted into the hydrofarm greenhouses. In another process at ADM, carbon dioxide is produced during the fermentation of alcohol. Since plants grow by breathing carbon dioxide instead of oxygen, carbon dioxide is moved from the fermenting process to the greenhouse, and the natural level of 340 parts per million is increased to 1,200 ppm. The plant yield can be increased by 20 percent.

The economics of hydrofarming are different from regular farming. The capital investment is high—the greenhouse and the chemistry are costly. The low-cost heat from the other operations is necessary to keep the crop competitive with shipped-in lettuce. The labor cost per head of lettuce is much lower than soil-grown lettuce, since more of the production can be automated—admirably so.

There are a number of other hydroponics

Boston lettuce at Archer Daniels Midland is grown hydroponically in a greenhouse heated with waste heat that has been generated elsewhere.

operations in the United States, but ADM's is by far the largest. There's no reason it couldn't be duplicated anywhere there is heat from generating or from industrial processes. Space requirements are minimal, though expensive.

Hydroponic production isn't going to take over agriculture, but it may well find a niche. Because it requires major "up front" capital investment, it will likely remain the province of large companies, although small-town heat sources such as power generating plants may encourage small hydroponic vegetable operations for local consumption.

Politics

If there is anything certain about agriculture's future, it is that it is bound up with political decisions. American agriculture simply can't be separated from American politics. The combined history is too long, the alternatives too frightening. The soaring cost of farm programs may eventually cause them to be cut back in cost, but it isn't possible to wipe out political control entirely. We'll wander back and forth within a relatively narrow middle ground, muddling through, never satisfactorily, but hopefully never disastrously.

I've never met anyone who can adequately explain the purpose, or adequately describe the successes or failures of our farm programs. The need for knowledge is too great: the ramifications of the programs turn out to be so unexpected, the political requirements so overwhelming. And yet, an attempt at understanding it is necessary because the situation determines the way of life for several million people, our source of food, not to mention the annual expenditure of about $25 billion of your money.

Let's look, first, at how the government, and hopefully the public, thinks of farmers, for that is basic to the entire problem. Second, let's look briefly at the techniques being used to carry out the farm policies, and what these policies are trying to do. Finally, let's look at alternatives.

Keep in mind that there is great disagreement on the entire subject of farm policy, disagreement broader than Democrat and Republican. Certainly not even all farmers think alike; in fact, few do. Worse, what they say they think is often not what they really think. Worse yet, sometimes they realize it and sometimes they don't. The same applies to congressmen and bureaucrats, and certainly to the public.

Nowhere is this confusing viewpoint more

The emotional effect of a farm auction is always powerful. Sometimes an auction simply means a farm family has decided to retire and move to town, but it can also mean the creditors have forced the farmer out.

true than in the farmer's oft-repeated complaint: "I just want to get the government out of the whole mess and off my back." In fact, not many farmers really want that. They want government out, it seems, but not out of the area in which government supports *them*. When this is pointed out, the farmers generally say, "Yeah, but government's been interfering so long I no longer have a choice." Some truth there.

Throughout the mid-1980s we've been exposed to numerous television dramas, movies, and news programs telling us that farming is dead, and the evil government or evil bankers have foreclosed on just about everyone. Things really aren't that tough, though. Farmers and farm publicists long ago discovered that the public has sympathy for the farmer, and that the squeaky wheel gets greased. The farm propaganda machine is powerful.

Is the farmer different from, say, the independent widget manufacturer? We know he is. The government doesn't buy up surplus widgets and stockpile them, avoids setting a low price to give the consumer a bargain, and never pays the manufacturer the difference between what the consumer pays and what the widget lobby convinces Congress the manufacturer and his employees need as income. If widgets are sold abroad I'm not aware the government subsidizes them. And if the manufacturer is going broke, the government usually doesn't lend him the money to keep going—Chrysler maybe, but widget manufacturers, never.

Sounds ridiculous, doesn't it? Widget makers go broke every day. The better producers stay in business, and they are presumably sensible enough to make just enough widgets to allow them to stay profitable if they manufacture the best widgets possible without flooding the market. Failing that, they find some other product—or they go broke.

For many years it has been the policy of our society, and therefore our federal government—or possibly the order is the reverse—not to accept farmers going broke. Sometimes farmers actually do go broke, but the blame is seldom put on the farmer; rather, the "system" is at fault.

In the eyes of the public is the farmer any different from the widget manufacturer? Yes, in many ways he is. Many of us have farmers as ancestors, few have widget manufacturers. Most of us love apple pie, and the farmer is more American than apple pie. Many of us think enviously of the farmer's independence as the American dream, although fortunately not many of us really want the effort or risk it takes to farm. We likely understand that farmland, in some mystical but not legal sense, belongs to all of us. Emotionally, we are somehow attached to farmland and to farmers.

Clearly, more of us have greater sympathy for the risks of farming than for the risks of manufacturing widgets. I'm not arguing that this is wrong, but I am pointing out that it exists and it affects our national view of farming.

One final, but vital, difference between widget manufacturers and farmers: when the widget manufacturer goes broke, his production generally ceases entirely. If the farmer is forced off the land, the land doesn't disappear, but generally continues to produce. If the farm owner goes broke, the land is taken over, usually by a more efficient, or at least a more financially astute neighbor. Barring

Many of the government programs are aimed toward land conservation, particularly in areas like this that have a minimum of precipitation. Under the relatively new Conservation Reserve Program, land that is dangerous to farm can be rented to the government for a ten year period. It is planned that there will be 45 million acres in the program, an area the size of North Dakota.

government edict or other persuasive incentives, there really isn't much land taken out of production, regardless of the original owner's condition. That is a basic difference between widget makers and farming.

The Past and Present Situation

Now let's take a look at how agriculture used to function, how it presently works, and the alternatives for the future.

Sometime in the dim past, agriculture functioned pretty much as other private markets function. When farm production got too great, the market became glutted and prices dropped. Production would decline, sometimes disastrously for both producer and consumer. Although we've long heard of the approaching time when we will all starve, it happens that for many years we've been able to produce more than we can get rid of. (Let's not get bogged down here with the starving third world nations; whether we feed them or not won't actually change our basic farm problems, although feeding them should certainly ease our consciences.)

By the Depression years of the 30s it was obvious our farmers had been producing more than consumers could buy, and the relatively free market competition had driven prices below the cost of production. Our farmers weren't making it. Even the disastrous droughts of the 30s didn't solve the production problems and restore prices to a livable level.

In 1933 the Roosevelt Administration made the first effort toward major manipulation of the market. The Agricultural Adjustment Act started to pay farmers for cutting their production acreage in the six main crops. We continue the practice, although we've added subtleties undreamed of by Roosevelt's Secretary of Agriculture, Henry Wallace. In succeeding farm bills we've tried to remedy the failures of earlier laws, sometimes succeeding, more often obfuscating the issues. But we've essentially continued the earlier ideas.

Basically, our major farm programs consist of three activities. Legislation is aimed at supporting farm prices, supporting farm income, and cutting production through acreage controls. The goal of these activities is to get prices and production in balance, so the consumer has fair prices and the farmer makes a fair living. The basic controls are directed at the major crops, but many additional crops are controlled through a wide variety of other techniques. The crops and animals that are not controlled generally seem to have muddled through independently, although in fairness they've sometimes encountered hard times.

The first activity, supporting farm prices, is designed to put a floor on crop prices. The government makes loans to farmers at prices decreed by government, using the farmer's crops as collateral. If market prices rise above the price the farmer has borrowed at, he can sell the crop on the market and pay off the loan and its interest, and keep the profit. If market prices drop lower than the loan, he can turn the crop over to the government in payment of his loan, and the government will then hold the crop and pay for its storage. In effect, the government has set a floor for the price of the crop. The farmer will always get paid the amount of his loan for the crop.

For rice and cotton there is a new wrinkle

to this called a marketing loan, the purpose of which is to encourage farmers to sell in the foreign market, where prices may be lower than in the United States. Under certain conditions, the cotton or rice farmer can sell abroad at the existing foreign market price, and pay off his government loan with what he gets, the government absorbing the difference between the foreign price and what it has lent the farmer. This technique has helped immensely in our becoming competitive in the foreign markets we had lost. Although the taxpayer pays for it, some of the surplus product is disposed of. At the same time, this method also spurs additional production, which is what we don't want.

The second goal of farm policy, supporting farm income, is designed to keep the farmer solvent. This income boost is handled through what is known as the "target price." Congress simply agrees (well, sometimes it isn't so simple) on the price farmers should be guaranteed for their major crops. The government then pays farmers the difference between the higher of either the loan rate or the market price, and the target price. For example, if the loan rate on corn is $1.75 a bushel and the market price is $2, and the target price has been set at $3, then the farmer gets the $1 per bushel difference between market and target, plus the $2 when he sells the corn at market price. He then pays back the $1.75 loan.

Obviously, with the cushion of price supports and target prices, the farmer would keep pushing production up and up, taking in more and more dollars, if he could plant as much as he wanted. For this reason, it's been necessary to create a stopper. Acreage controls for the major crops are the stopper, an effort to reduce production by decreasing the amount of land under cultivation. The farmer is free to produce as much as possible, but only on an *allocated* number of acres. The acres he can plant are generally established as a percentage of his stated "base acreage," determined by his historic production. In 1987 wheat growers could plant only 72.5 percent of their wheat base, corn growers could plant 80 percent of their corn base. The acres not planted are called diversion, or set-aside, acres, and they can't be used for any harvestable crop.

This set-aside ground is part of the Acreage Conservation Reserve. Since the farmer can choose what ground goes into the ACR, he naturally selects his poorest fields, which means that if he has to idle 20 percent of his ground his production goes down less—his better ground produces more.

Another production-cutting technique has increasingly appealed to conservation needs. This method means killing two birds with one stone: cutting back on production by aiding conservation at the same time is really like being against sin: it's hard to beat. Land currently in crop production that presents serious erosion problems can be retired for pay in various ways for long periods. The Conservation Reserve Program is aimed at getting 45 million acres back into grass by paying the owners an annual rental price for ten years. Its cost is estimated at between $1 billion and $2 billion annually.

While these programs are worth every nickel they cost us, we should recognize that we are tackling the production problem obliquely. The land that goes into set-aside

Once a source of farm power, windmills have been largely displaced by energy based on petroleum or electricity. Just as windmills have been phased out, so has manpower. Haymaking has decreasingly relied on muscle power, increasingly on mechanical power. Efforts to return to wind or solar power during recent energy crises have generally had little effect. Increasingly, we will be dependent on the economies of mechanization.

While adding some of the conveniences of city living, farm life has retained many of the nostalgic pleasures it has always had. The pace is faster, the freedom possibly less, but for many, farming is still a wonderful way of life. Although full-time farming is unfortunately available to fewer and fewer families, the number of small-scale part-time farmers is increasing, and their lifestyle can have many of the pleasures of traditional farm life.

and other conservation programs is generally the land that produced the poorest crops to begin with.

The Result

Price support, income support, and acreage controls are the major farm programs aimed at bringing production into balance with market needs—without bankrupting farmers. There are niceties, complexities, and ramifications almost beyond calculation—after all, we've had fifty years to complicate them, and farmers have had fifty years to figure their way around them, which they've done creatively. Dozens of other crops are now involved, often in programs very different from the major programs, usually with much less government money at stake—a few million here a billion there.

A rundown of the names of some programs is available in the Agriculture Stabilization and Conservation Service records of payments to producers. Keep in mind that legislators have gotten adept at designating names that are not really descriptive—such as calling the agricultural bill of 1985 the Food Security Act of 1985.

Here are some of the programs: Cotton Disaster Program, Cotton Deficiency Program, Cotton Voluntary Diversion, Cotton Loan Deficiency, Feed Grain Disaster Program, Feed Grain Deficiency Program, Feed Grain Voluntary Diversion, Rice Deficiency Program, Rice Voluntary Diversion, Rice Inventory Program, Rice Marketing Expense Program, Wheat Disaster Program, Wheat Deficiency Program, Wheat Voluntary Diversion, National Wool Act Program, Agricultural Conservation Program, Forestry Incentive Program, Emergency Conservation Program, Water Bank Program, Emergency Feed Program, Extended Storage Program, Extended Warehouse Storage, PIK Storage, Rural Clean Water Program, Clean Lakes Program, Conservation Reserve Program, Animal Waste Management Program, Milk Diversion, Dairy Indemnity Program, Dairy Termination Program.

This is by no means a complete list of the ways farmers get paid—in 1986 the above programs came only to about $12 billion, while the entire farm program costs amounted to $20 to $30 billion. There's mohair to be accounted for and I think there's support for honey, too; there's a Great Plains Conservation Program for about $10 million, and some good new ones are coming along. The bureaucracy may be incredible, but it is apparently tolerable.

A caveat. It's true that we're looking at a lot of money here. As with all federal expenditures, however, we need to look at the big picture. In 1987, USDA expenditures for farm programs came to about $20 to $30 billion—give or take a few billion, and subject to a lot of definitions. Twenty-five billion dollars amounts to only a little more than $100 for each of us to ante up. Our federal deficit runs ten times that much; the defense budget comes to more than ten times the cost of farm programs.

Alternatives

New alternatives are constantly being offered. As costs of existing programs increase, or as failures become more evident, alternative proposals proliferate. At one extreme, there was mumbling from the Reagan Administration that programs would be phased out entirely and a free market reinstalled. In somewhat the same vein, there have been proposals to "decouple." Essentially, decoupling would pay the farmer a wage, usually proposed as a "decreasing" wage, over five years, perhaps, during which time he would be allowed to produce whatever he wanted and sell it at whatever he could get. I look at it as welfare (never mention that word to a farmer) on a declining scale, an incentive to get out and do something else. These ideas are pretty drastic, probably too drastic to be adopted.

At the other extreme is an attempt to face up to the reality of controls by making them work more directly. One such plan has been put forward but not adopted as of this writing. This is the Harkin-Gephardt "Save the Family Farm" act that would, very simply, put mandatory controls on actual bushels or pounds being produced, rather than on the acreage. A quota system on the main crops would be instituted. The purpose would be to limit output, thereby raising prices and allowing subsidies to be decreased. Production quotas would be set for each major crop, just as acreage controls are now established, but in actual quantities. Quotas could probably be bought and sold among farmers. Although I doubt the proposal has much to do with "Saving the Family Farm," it is a different approach to the major crops, even if a similar program has functioned for years in the minor crops of peanuts and tobacco. Again, it is probably too drastic to be acceptable.

The Future

That leaves us with entirely new ideas, as yet unknown, or with continued muddling through. I'll bet on muddling. We've been at this game of price supports and/or production manipulation for more than fifty years, and both politicians and farmers have become pretty good at it. As federal programs have proliferated, farmers have discovered ways to get program money but work their way around compliance. Programs have been piled on programs; bureaucracy has settled in, while slightly different approaches have been tried; some proposals have been abandoned, some have simply been made more complicated and renamed.

Meanwhile, the problem certainly hasn't been solved, although it has had some success. Our programs have been expensive and unsatisfactory, but I frankly don't have a better plan. Most of the obvious small improvements are not politically feasible. Present programs have generally kept the consumer satisfied and the farmer, if not rich, at least not bankrupt.

For all this, what do we get? Is our food cheap? Supposedly we spend less of our income on food than people in any other country. I'd bet, though, that the prices for food would be a lot lower if the government just said, "All bets are off, boys, you're on your own. Let 'er go and we'll see where she lands!" Don't wait for that to happen, since it would mean disaster for our entire economy.

Are we producing a feeling of security among farmers for our billions? Obviously not, although we've staved off a lot of farm bankruptcies, sure enough. We've managed over the years to "gentle" the needed drop in farm numbers, but we are still having trouble decreasing the amount of land in production to meet our needs. For a while we thought the world's food needs would catch up with our production, but that seems unlikely in the forseeable future. We've got to get on with lowering production, gently, expensively, but get on with it, maybe even more harshly than we have. I believe this requires forcing more farmers off the farm, disturbing as that may sound. It has to be done gently, with as little trauma as possible, but done.

Finally, accepting the historical ties between farming and politics as unavoidable, I think we're heading generally in the right direction, staggering back and forth along the way, putting up with a ridiculous amount of bureaucracy, getting cheated by farmers who don't need help in the first place and have learned to beat the system. It's all askew and confused, and it certainly would be better if it were neat and clearheaded. But at this point we're too immersed in government finagling to cut loose and start over. We'll just have to muddle through, hopefully making little gains while avoiding disaster, succeeding more than failing.

Acknowledgments

First, let me thank all the patient people who allowed me to interview them. You'll find their names scattered through the book. No one I asked turned me down, and interviews are time-consuming for a lot of people whose time is important.

Second, there are dozens of members of the bureaucracy who were helpful. Within the USDA there are the ERS, FAS, ASCS, Research, Statistics, FLB, CCC, FCA, extension offices throughout the United States, and more; over in Commerce there is the Agricultural Census. Most states have their own panoply of statisticians and agricultural relations people. The statistics are not so difficult to get, but it's the explanations behind the statistics that these people are so expert at.

Conversations with Department of Agriculture specialists were likely to go like this:

"I need information on the number of pounds of bread Russians consume in a year, compared to U.S. consumption. Can you help me?"

"Good question, that's not my special field. I work only on Russian grain production, not who eats it. But Mary is right here, she's doing a report about that."

"Oh, Mr. Heilman, I know the figures you have. Don't believe them; it's not that simple. You see, Russian bread is kept at a very very low price so Russians can tell their citizens how cheap their bread is. It's so cheap that instead of buying grain for their private animals, the citizens simply feed them bread. That skews the figures . . ."

It's always astonishing to me to find individuals who deal daily with some tiny but often vital set of statistics. They have been not only accommodating but eager to share the information and to explain what the figures really mean. Sometimes I've been able to understand.

There are two important government publications in the agricultural field. The 550 action-packed pages of *Agricultural Statistics* are the basic source of most facts. The *Census of Agriculture* is the other major source. The latter, unfortunately, is conducted only every five years, and the latest now available dates from 1982. Still, they are both invaluable to any researcher, although I don't recommend them as bedtime reading matter.

Farm trade organizations, of which there are many dozens, are an additional source I've used. Some are the publicity arms of a particular crop or animal. Others are much more important for statistical information. Some are both. A former farm magazine editor, Paul Weller, has been the source of who knows what; he's agriculture's champion public relations man, executive secretary of more organizations than I could ever even belong to.

Also in Washington there's a retired USDA public relations person, Ed Curran, who privately publishes *The Farm Paper Letter*. I've known him the past thirty years, and he and *The Farm Paper Letter* have been an invaluable source for the book. He knows everyone, pounds the pavement for the latest facts and gossip, and is able to separate the two.

And finally, Susan Meyer, Marsha Melnick, Meg Ross, and Virginia Croft at Roundtable Press, who edited, cajoled and advised, and were responsible for my doing the book in the first place. And not only are we all still speaking, but we are friends.

Errors are, as usual, mine.

Index

Page numbers in italics refer to illustrations.

Picture Credits

All photographs are by Grant Heilman
    except for those by the following
    photographers:
Robert Barclay, page 184
John Colwell, pages 175, 177, 179
Isaac Geib, page 191
Thomas Hovland, pages 178, 182
Larry Lefever, pages 28, 29, 31, 77, 94, 99,
    108, 132, 135, 136, 149, 156, 170, 173,
    179, 207, 226
Gary McMichael, page 147
Alan Pitcairn, page 42
Barry Runk, pages 214, 215
Runk/Schoenberger, pages 60, 117, 119,
    156, 214, 230, 231, 258